CHICKEN FACT OR CHICKEN POOP

Brimming with creative inspiration, how-to projects, and useful information to enrich your everyday life, Quarto Knows is a favorite destination for those pursuing their interests and passions. Visit our site and dig deeper with our books into your area of interest: Quarto Creates, Quarto Cooks, Quarto Homes, Quarto Lives, Quarto Drives, Quarto Explores, Quarto Gifts, or Quarto Kids.

First Published in 2018 by Quarry Books, an imprint of The Quarto Group,
100 Cummings Center, Suite 265-D, Beverly, MA 01915, USA.
Phone: 978.282.9590 Fax: 978.283.2742 QuartoKnows.com

Quarry Books titles are also available at discount for retail, wholesale, promotional, and bulk purchase. For details, contact the Special Sales Manager by email at specialsales@quarto.com or by mail at The Quarto Group, Attn: Special Sales Manager, 401 Second Avenue North, Suite 310, Minneapolis, MN 55401, USA.

10 9 8 7 6 5 4 3 2 1

ISBN: 978-1-63159-315-4

Digital edition published in 2018

Library of Congress Cataloging-in-Publication Data available.

Cover Design: Landers Miller Design
Page Layout: Landers Miller Design
Photography: Shutterstock.com

Printed in China

THE CHICKEN WHISPERER S GUIDE

CHICKEN FACT OR CHICKEN POOP

TO THE

FACTS AND FICTIONS

YOU NEED TO KNOW TO KEEP YOUR FLOCK HEALTHY AND HAPPY

—ANDY SCHNEIDER—
THE CHICKEN WHISPERER

"Show me the proof."

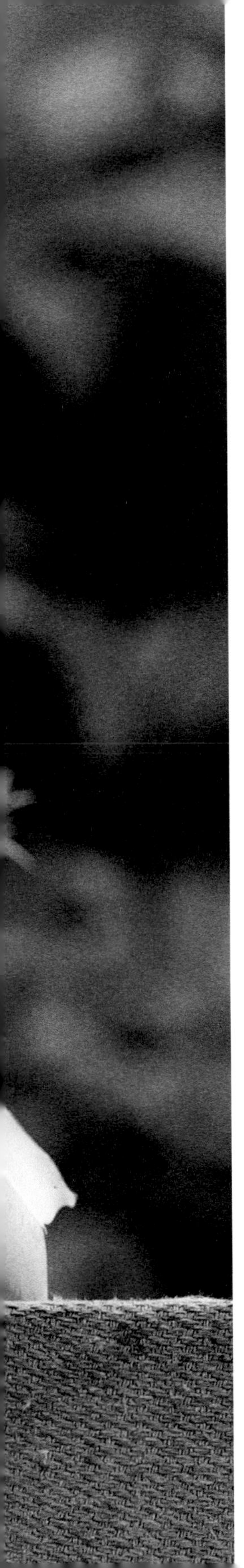

Preface

In every workshop I teach, during the Blog Busters segment, I tell attendees that their four favorite words when reading chicken blogs and chicken forums should be "show me the proof" and if the only proof they receive is, "It worked for me," they should tread lightly with the posted claim. About two years ago, I got so frustrated with all the bad information being shared on chicken blogs and chicken forums that I came up with the idea for a Fact or Chicken Poop website. It was such a success, I thought there was a great opportunity for the release of my second book based on the website. So, why have a book and website solely based on calling out the bad information? Because there is so much bad information out there, something had to be done. Social media, like so many other things, can have both good and bad outcomes. Unfortunately, the bad information would be shared, and some of the information would go viral.

So, where does the bad information come from? Most of the time, it's just shared . . . and shared and shared. But where did it originate? Most of the time, the information comes from folks that claimed they have done or used a certain method or treatment and claimed it worked. They had no proof it worked; they just claimed it worked. Here's a classic example of this: In the fall, people share that pumpkin seeds are an all-natural dewormer for chickens. When I see this post year after year, I always respond the same way by asking the person who posted it to show me the proof. They will in turn reply with, "Well, I give my chickens pumpkin seeds, and they don't have worms." I then ask if they have ever actually had their chickens tested for worms by a veterinarian, and the answer is always no. Therefore, they truly have no proof whatsoever to back up their claim.

In most cases, they just saw this information posted on a chicken blog or chicken forum and are now spreading bad information. I did, however, have a young woman that thought she had stumped me and stumped me good during one of these discussions. When she claimed that pumpkin seeds were an all-natural dewormer for chickens, I went down the same path I always do. I asked

"Unfortunately, the bad information would be shared, and some of the information would go viral."

her if she would mind sharing some proof to back up her claim. She responded like the others: She gives pumpkin seeds to her chickens, and they don't have worms. I asked her if she had ever actually had her chickens tested for worms by a veterinarian, and she responded, "Yes, I have, and my chickens don't have worms." She thought for a second that she stumped me. However, I responded with asking her, "So your chickens at one point actually *had* worms? And she replied that her chickens have never tested positive for worms. So, you can see where this is going. I then replied, "So how do you know that pumpkin seeds are an all-natural dewormer if your chickens have never had worms to begin with? In order to make your claim that pumpkin seeds are an all-natural dewormer for chickens, you need to start with chickens that actually tested positive to have worms. Then, I want to know what variety (type) of pumpkin seed you gave them, how much (dose) pumpkin seeds you gave them, how long you administered the dose of pumpkin seeds and how you made sure that every chicken received the same dose of pumpkin seeds, and finally, a test that showed they no longer had any worms!" This is a perfect reason why this book exists!

Sometimes, the information comes from a study that was misread or misunderstood. A classic example of this is when a chicken blogger claimed that giving baby chicks probiotics would reduce the risk of *Salmonella* later in life. She referred to a study that was conducted on broilers, not layers. So now that this blogger has spread bad and incorrect information, there is no telling how many people are going to believe this nonsense!

There are even so-called all-natural chicken keeping blogs and websites that sell several chicken elixirs that claim to cure, prevent, and treat just about anything your chickens could have but, of course, don't offer any scientific data or proof those elixirs work other than the four words I discussed above, "It worked for me." And even that response is not accurate because they have no idea if it actually worked!

It's bad enough that the chicken blogs and chicken forums are loaded with this bad information, but wait, there's more! A couple of years ago, a chicken-related television show was launched. I was actually invited to appear on one of the episodes and even taped my entire segment before the first episode ever aired. Once I started to see some of the information that was being shared on other episodes, I contacted the producer and immediately canceled my appearance in the episode. Why? I asked the producer if anything aired was actually being fact-checked for accuracy. His answer? No. So there was no way I could be associated with this show. And even today, I still see information shared by this chicken-related television show that is nothing more than quackery! While I'm sure these people have good intentions, they are actually hurting the backyard chicken movement by offering questionable, marginal, and sometimes outright harmful information.

Lastly, I know this book, my other book, my magazine, and my social media pages like Facebook and Twitter are not for everyone. Some folks don't want the truth. They don't want science-based, fact-based, and study-based information. They want to be told what they are doing is right and refuse to accept anything else, especially if it proves what they are doing is wrong or harmful. The Chicken Whisperer is no longer about a person. It's become a brand of sorts, and when people see that brand, they think of science-based, fact-based, and study-based information they can count on to raise a healthy flock of chickens. Many have told me they still visit the chicken blogs and chicken forums to see funny pictures of chickens, chicken jokes, chicken crafts, and chicken-related treats. But they also tell me when they want science-based, fact-based, and study-based information—the right information to provide care for their flock—they visit the Chicken Whisperer to get that information.

—**ANDY SCHNEIDER**
The Chicken Whisperer

1

Home Sweet Home: Your Life with Chickens

"Adding red pepper flakes to your chickens' feed will increase egg production."

☐ FACT ☐ POOP

TOPIC

HOME SWEET HOME NO. 1

ANSWER PROVIDED BY

Maurice Pitesky, DVM, MPVM, DACVPM
Veterinarian/Assistant Specialist
in Cooperative Extension Poultry Health
and Food Safety Epidemiology
School of Veterinary Medicine
University of California, Davis

The research

From a generic perspective, feed is necessary for laying hens to provide energy for body maintenance, growth, feather production, and egg production. More specifically, only sexually mature hens can anatomically produce eggs. Furthermore, most laying breeds need approximately fourteen to sixteen hours of light a day to remain in lay. Changes in lighting exposure, intensity, and health can moderately to severely affect egg production.

Nutritionally, birds in lay need a calcium-rich diet (i.e., dietary calcium between 3.25 percent and 4.5 percent) with a relatively high level of energy. Because eggs intuitively take a significant amount of energy to make and calcium carbonate is the most common nutrient in eggshells, the nutritional need for energy and calcium should make sense. Though science has well-defined desired calcium levels (with some variability based on genetics), energy requirements can vary based on genetics, temperature, and percentage of feather cover. For example, hens in cold temperatures with inadequate feather coverage need to consume additional feed for body maintenance, growth, feather production, and egg production. Husbandry practices, therefore, can affect production.

With respect to red pepper flakes, we're talking here about dried chilies, which can include serrano, jalapeño, ancho, and bell peppers. Spiciness equates to the Scoville heat unit, which measures the concentration of the chemical compound capsaicin. This active chemical produces the heat sensation on our taste buds. Interestingly, birds do not have receptors for capsaicin. The reason probably relates to evolution. When birds consume peppers, the seeds pass through the gut undigested and then are deposited in new locations where they can propagate. If birds were sensitive to capsaicin, they would not eat many types of spicy peppers and hence would not spread those seeds as willingly.

The verdict

The answer to this is 100 percent poop. Based on the above and a thorough review of the scientific literature, it is difficult to understand any potential mode of action that peppers would supply to birds to increase egg production—either size or rate.

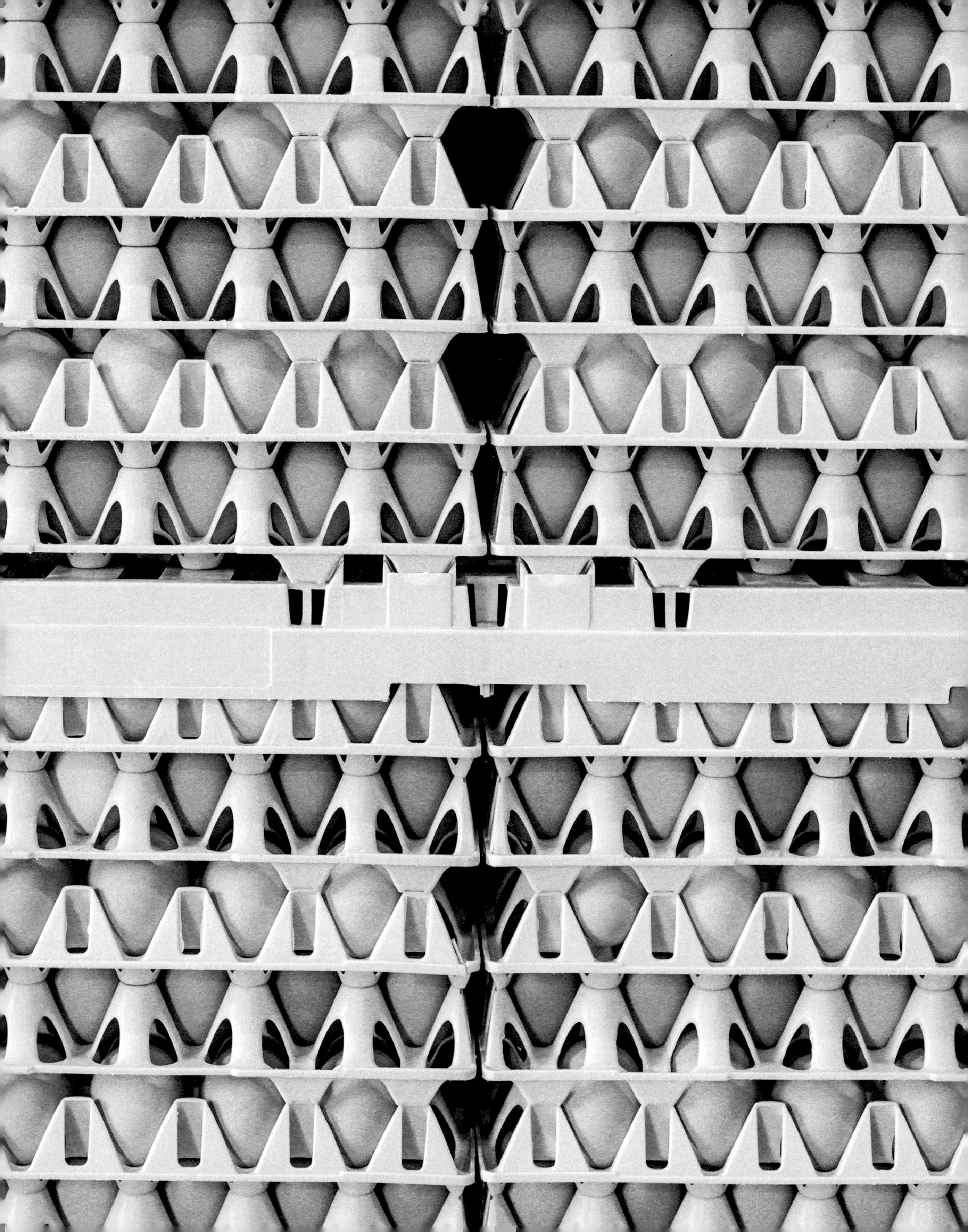

“Eggs produced at home are actually more expensive than eggs you buy at the store.”

☐ FACT ☐ POOP

TOPIC

HOME SWEET HOME NO. 2

ANSWER PROVIDED BY

Brigid A. McCrea, PhD
Extension Specialist
4-H Youth Development Animal Programs
Alabama Cooperative Extension System
Auburn University

The research

Eggs are a complete and nutritious source of protein. There are a number of variables that govern the affordability of the eggs that make it to the grocery store: the season, the age of the flock, the price of gasoline to transport the eggs, and the overall number of chickens that are out there and available to lay those yummy eggs.

The overhead costs for commercial poultry producers means that they pay for the land, building, and labor for the flock. This is not unlike what you, the small flock owner, does when starting your flock. Hopefully, you are already in ownership of the land and have the permits in place to begin the construction of the coop. Building permits or poultry ownership licenses vary in every locale and in every state. It is up to you to make sure that you are in compliance and have all your paperwork in order. That usually costs time and money on your part. Those costs need to be factored into the price of your eggs. Let's say, for the sake of this example, that these costs—your time and the actual license—will be $35.

The cost of the chicks themselves depends on the number that you get. Many small flock owners have ten or fewer, so let's go with six birds. If you estimate that each chick costs $5 each, then that would cover the cost of vaccinations, sexing, shipping, and the actual cost of the chick itself. So you now have $30 invested in your chicks. Try not to let any of them die in the brooding and rearing process!

Brooding kits do exist, but a simple box and brood lamp in a spare bedroom can do the trick for the first few weeks. Then, you will want to move them to the coop to finish brooding in some cases. Brooding will run you approximately $50 if you clean regularly and if you include the cost of electricity and your labor in caring for the chicks. That also includes new equipment and a medicated starter diet for your chicks.

Then, there is the cost of supplies for building the coop. If you are not building your coop, then you will need to purchase a preconstructed coop. Now, this is when things tend to get rather pricey. Let's say that you construct the covered run for the coop, but the coop itself is prefabricated. Don't forget to include labor here. Chicken coop runs do not magically appear on their own! A conservative estimate for a flock of six chickens would be $1,500 for a prefabricated coop and then another $350 for wood, wire, renting a trencher to bury the wire, labor, shingles, and hardware for a door. Let's not forget the time and labor part of the equation as well. So, you may be looking at $1,850 or even more for a coop that is sufficient for your flock size.

Now, let's talk about feeding and care for your flock—the second priciest portion of chicken keeping. If you have six chickens, then they will go through a 50-pound (23 kg) bag of feed every month depending on the weather and their stage of production. You do not want them to waste that feed, so you purchase a feeder that is adjustable in height. You do the same with the waterer, choosing nipple drinkers to avoid spillage. Your feeding and watering equipment will likely run you $80 but is a worthy investment in your flock as long as you maintain it well. Your feed will likely run from $13 to $19 for each 50-pound (23 kg) bag. That is for regular feed, and we are not even going to delve into the costs for organic, GMO-free, or other specialty feeds that are more costly. Let's say you pay $16 for a bag of feed and that you go through one-quarter pound (23 kg) of feed per bird per day. That means you will go through twelve bags in a year, costing you $192. This cost is only in reference to the layer diet that they will receive when they are in the egg production stage of their life.

We cannot forget the growth stage. They will consume less feed at this time but will increase their consumption as they age. This grower feed costs about the same as a layer diet, so let's say $16 for a 50-pound bag. They will go through about four bags as they grow, so that cost will be $64. As you may recall, the brooding section covered the starter diet. For the grower and layer diets, you will have spent $256. Now, you may understand why so many people try to scrimp with chicken feed to save on their costs.

Bedding is another cost that many people initially forget. Bales of pine shavings will run about $6.50 each. Depending on the size of your coop, one bale

"Chickens are good pets for young children."

ANSWER PROVIDED BY

Megin Nichols, DVM, MPH, DACVPM,
Enteric Zoonoses Activity Lead Centers for Disease Control and Prevention

Chicken poop. Finding the right pet for your family is an important decision based on many factors. Adult chickens and baby chicks are inappropriate pets for children under the age of five because chickens are a common source of bacterial illnesses such as *Salmonella*. These germs can be passed to people directly through touching, cuddling, or kissing an infected chicken or indirectly through contact with the chicken's droppings, feathers, cages, or materials from its living environment (soil, hay, plants, feeding/water dishes, etc.). Young children are at a higher risk for infection, and in particular a more severe infection, as a result of their underdeveloped immune systems and their tendency to put their fingers or other items in their mouth. Severe infections in infants and young children can lead to death without immediate medical care.

after each cleanup may not be enough. So, for the coop described above, one bale is sufficient for a month or more if the feces are spot cleaned and picked up daily. So, if the bedding is removed and replaced monthly, then the cost would be $78 for a year's time. This also includes that brooding for six weeks has already included the cost of a bag of shavings that will last that whole time given weekly cleanings.

As you can see, starting a flock can be costly if you add in all the startup costs and the costs of maintaining a flock for a year after they have started laying. Your grand total is $2,379, so let's round up to $2,380 for an easier number to work with. If your average hen lays well in its first year and you get 250 eggs, then you would be really pleased. Multiply that by six hens and you will have 1,500 eggs in a year. Each egg will cost you $1.59 if you divide your costs by the number of eggs. Your dozen eggs from home will cost you $19.08. The average price for a dozen eggs at the grocery store is from $2 to $4, depending on market variables and where you live. Still, the discrepancy is pretty large when you look at the price for a dozen eggs from a backyard flock and those from the grocery store.

The verdict

Fact. Eggs are known as nature's perfect protein. It is the protein against which all other proteins are measured. And if you really think about it, an egg, if fertilized, can produce a whole new life because everything it needs is already inside. The fact that eggs are so affordable at the grocery store means that a nutritious source of protein is available to more people whose income may not allow them to raise their own flock.

"Sand is a wonderful surface for both runs and coops."

☐ FACT ☐ POOP

TOPIC

HOME SWEET HOME NO. 3

ANSWER PROVIDED BY

Brigid A. McCrea, PhD
Extension Specialist
4-H Youth Development Animal Programs
Alabama Cooperative Extension System
Auburn University

The research

For many years, sand has been called an excellent bedding material; even commercial producers have considered it. But brooding temperatures, weight, nutrient buildup, bacterial loads, and cost all remain factors to think about when comparing sand to pine shavings.

When putting down new sand or new pine shavings, brooding becomes a concern. A study published in *The Journal of Applied Poultry Research* (Bowers et al., 2003) showed that new sand is about 10°F (-12°C) cooler at maximum brooding temperature than pine shavings. Chilled chicks do not do well and may not develop as well as their counterparts kept at warmer temperatures. Researchers found that broiler chickens raised on sand versus pine shavings did equally well (Bilgili et al., 1999, *The Journal of Applied Poultry Research*).

Nutrients build up over time for all used litter, regardless of original litter material. However, sand does not retain them as well. So, if you are spreading composted or used litter on the ground later, then this may not be the right substrate for your soil.

Over time, sand as a litter becomes quite heavy, making it a challenge to remove and replace. Beware that when you put sand outdoors in the run, you may get low areas, so tamp it down and add more to create an even surface. Picking up droppings is easiest with a scooper. Sand inside the coop is not advised as it can provide a rough landing surface for the chickens when they jump down from the perch or nest box. Also, in warmer and more humid climates, ammonia levels over time could creep higher with sand bedding than wood shavings.

Finally, because sand or pine shavings will be the landing pad for fecal material from your flock, it's important to know that sand is the clear winner with regard to bacteria retention (Macklin et al., 2005, *The Journal of Applied Poultry Research*). Researchers found that sand had lower water activity, bacterial counts, and moisture levels.

The verdict

Mainly poop. Just for the brooding period, you may consider pine shavings over sand—if you have a suboptimal brooding environment. However, used sand, with organic waste material mixed in, does not stay cold. If you are brooding chicks on sand that has had enough time to rest before the next batch of chicks is placed on it, research has indicated that the maximum temperature of the used sand litter does not differ from that of used pine shavings. In the run, tamp down sand to create an even surface; in the coop, create a barrier between the wooden floor and the sand. With regard to bacteria, sand is the best bet.

What it means

The cost of materials for either sand or wood shavings varies slightly per region. The cost of sand will be higher up front, but will end up slightly less in the long run. Flock owners who commit to sand are in the game for at least two years before they take out and replace their litter. Pine shavings are available at almost all feed stores, so availability makes them easy to replace. A bale costs $5 to $10 each, and it's easy to clean up and replace with greater regularity than that of sand.

"Pointy egg = rooster. Round egg = hen."

☐ FACT ☐ POOP

TOPIC

HOME SWEET HOME NO. 4

ANSWER PROVIDED BY

Brigid A. McCrea, PhD
Extension Specialist
4-H Youth Development Animal Programs
Alabama Cooperative Extension System
Auburn University

The research

There is a very popular myth circulating among backyard chicken keepers at this time. It revolves around a research article that seems to indicate a very slight correlation between the shape of an egg and the sex of the chick inside. The sample size in this research trial was very small. The lack of robust numbers in this research leaves ample room for questions.

In 1924, it was proven that the shape of the egg, called the shape index, had no bearing on the sex of chicks in domestic fowl. This research was done with thousands of eggs. One of the first things that a researcher examines in assessing the appropriateness of a journal article is the rigor of the research and sample size. If either of those things is lacking, then authors, at times, are able to publish in a peer-reviewed journal with a lower impact factor.

Another option for an author is to seek publication in a regionally specific journal in order to support its further growth and development. Perhaps such is the case of the publication in the Brazilian journal by the Turkish researchers. Either way, the number of eggs, and the use of just one specific strain, leaves one wondering if the results would be different with more samples.

Similar questions about the shapes of other bird species and the sex of chicks has also been asked by researchers. Just because it has been proven false in one species may not necessarily mean that it is true for all bird species. The trouble with research with wild bird species involves the interference of researchers with the nest and nestlings. Researchers must conduct their research in such a manner that it does not unduly cause the death of a hatchling or abandonment of a nest by the parents.

The verdict

Poop. The question of the shape or weight of an egg as a sign of the sex of the chick within has been asked many times for many species. Things to consider are the incubation period, the conditions under which the egg incubates, and the age of the hen that has laid the egg. Another study from Turkey, one with a larger sample size, found no relationship between the shape index of the egg

and the sex of the chick that hatches. As it stands with chickens, research has indicated that there is no correlation between the shape of the egg and the sex of the chick inside.

"Clean the manure every day from your coop."

☐ FACT ☐ POOP

TOPIC

HOME SWEET HOME NO. 5

ANSWER PROVIDED BY

Brigid A. McCrea, PhD
Extension Specialist
4-H Youth Development Animal Programs
Alabama Cooperative Extension System
Auburn University

The research

Everybody has a slightly different management scheme when it comes to both caring for and enjoying their chickens. Everyone knows, however, that chickens generate manure several times a day. So, how often should you be picking up the droppings from your chickens anyhow?

Well, let's start with the brooder. Depending on the number of chicks you are brooding, you may be able to go a couple of weeks before cleaning. But do not expect to get away with that for much longer than a month. In cases where you have a big plastic storage bin and maybe three or four chicks, you can get away with this scheme.

If you have a larger group—say, ten or more chicks—then you are going to clean that brooder out weekly, or else you and your flock will pay the price! What price, you ask? The smell of ammonia will permeate your home if you wait long enough!

For the rest of the lifetime of your flock, you can either pick up droppings daily off a droppings board under the roost or do a full litter cleanout weekly. It is up to you. Many people find that spending five minutes once a day on cleaning off a droppings board is worth the effort. Also, if you use a scooper and a bucket for the rest of the coop and run, the cleanup is a snap! Take that bucket right over to your compost pile and you will eventually see your compost grow into a nice pile.

Did you know that the average large fowl hen will produce about 130 pounds (59 kg) of manure in a year? Multiply that by the number of hens that you have in your flock and you can see that you will have a tidy sum of manure, given time. That means that the average hen, on a daily basis, produces just over one-quarter pound (115 g) of manure. However, remember that the initial weight is mostly due to water, which means that the manure will quickly dry out.

So whether you are picking wet or dry manure, you can count on a regular supply. Most small flock owners that use droppings boards say that they like to sprinkle sawdust over the surface to make cleaning go much faster. The sawdust keeps a small layer between the painted board and the droppings so that the droppings will not stick. That way, the owner can then just use a cat litter pan scooper to pick up the droppings and shake of the excess sawdust back onto the board. Of course, a little bit of sawdust is lost every day, but it can be easily replaced by sprinkling more on the board.

Then comes the question of using sand as litter. Sand is easy to find and apply, but keep in mind that it is very heavy. If you should need to replace the sand litter, then know that it only gets heavier with time. Sand is also not a good choice for brooding chicks because it is cold and chicks need warmth for a good start.

The verdict

Mainly poop. There is just not a lot of information out there regarding deep litter and scientific research. One of the concerns is if a bevy of bacteria develop within the deep litter that affect the overall health of your flock and the safety of the eggs. In such a case, owners would need to consider switching out of the low-input deep litter method to perhaps one mentioned above. In any case, deep litter does not mean that your flock should be left with rotting food or kitchen waste at any time. Pick up such items before they begin to decompose.

The moral of the story is to use whatever cleaning regimen keeps your personal flock their absolute healthiest without driving you crazy!

“Backyard poultry are different than commerical poultry.”

ANSWER PROVIDED BY

Megin Nichols, DVM, MPH, DACVPM

Enteric Zoonoses Activity Lead Centers for Disease Control and Prevention

It depends. “Backyard poultry” is a term used to describe a smaller flock of poultry, typically located at a private residence for personal use for eggs or meat. Backyard poultry might also provide chicken meat and eggs to local food systems, such as farmers’ markets. “Commercial poultry” is a term used to describe poultry in large-scale food production systems, such as those that produce chicken meat or eggs for distribution across the state or country. Backyard poultry owners obtain their poultry from mail-order hatcheries and/or agricultural feed stores. However, for certain poultry breeds, mail-order hatcheries may source poultry from the commercial poultry industry.

"Roosters can't see at night."

☐ FACT ☐ POOP

TOPIC

HOME SWEET HOME NO. 6

ANSWER PROVIDED BY

Maurice Pitesky, DVM, MPVM, DACVPM
Veterinarian/Assistant Specialist in Cooperative Extension Poultry Health and Food Safety Epidemiology
School of Veterinary Medicine
University of California, Davis

The research

While there are anatomical differences between males and females, eyes are anatomically and functionally similar.

Actually, avian eyes are truly remarkable. Birds can recognize constellations for stellar orientation and navigation and have high resolving power to perceive rapid movement. Some studies even suggest that migratory birds can "see" or perceive magnetic fields that help them navigate.

In addition, birds are capable of color discrimination, pattern recognition, and vision in bright and dim light, underwater accommodation, and sun orientation. They can also see in the regular and ultraviolet spectra. In short, vision is the most important sense for birds, and their vision is better than that of humans.

The verdict

Poop. Roosters can see well at night and equally as well as hens. It's possible this idea derived from comparing younger hens to older roosters. Roosters can be useful for longer than laying hens, so over time, older roosters may not appear to see as well as their younger, female counterparts.

"Raw chicken manure can be put in your backyard without composting."

☐ FACT ☐ POOP

TOPIC

HOME SWEET HOME NO. 7

ANSWER PROVIDED BY

Brigid A. McCrea, PhD
Extension Specialist
4-H Youth Development Animal Programs
Alabama Cooperative Extension System
Auburn University

The research

Chicken manure is well known to be a great amendment to your garden. The N-P-K or nitrogen-phosphorous-potassium ratio of chicken manure tends to be well balanced for use in the average homeowner's vegetable garden. In general, there are three concerns with applying fresh manure to a garden.

First, fresh chicken manure is mostly water. In fact, water constitutes about 76 percent of the weight of fresh manure. So a lot of what is lost or leached out right after fresh manure is applied to any location, be it a garden or a compost pile, is water. As manure ages, not only is water lost, but the nutrient composition changes as well. Some nutrients are released as a gas or convert to forms that are either more or less available to the plant. One prime example of this is the nitrogen content of fresh chicken manure. Nitrogen comes in the form of uric acid and ammonia. The uric acid is a solid that breaks down into urea and then breaks down into ammonia gas. From there, that ammonia gas is released into the environment.

Second, applying fresh manure to the garden has the potential to be "too hot" for the plants. The terms "hot" or "burn" are in reference to causing the plant harm. If you work fresh manure into the ground and transplant too soon afterward, then the manure will have burned the crop and can cause you to lose the transplants. You can also harm your plant if you put manure too close to larger plants or if you incorporate too high of a percentage of manure into your soil when preparing the soil for planting. If your chickens defecate in the garden in a regular location, such as near a certain plant or near the entrance, then the potential does exist for those plants to be on the receiving end of too many nutrients. If only we could see where the soil in our gardens needs improvement and could just send the chickens over there to do their business!

Third, and most important, is the concern about food safety. Commercial vegetable crop farmers learn about timing when to apply certain materials to the ground to avoid making the consumers of their final products ill. That does not mean that mistakes are not made. But it also means that you will need to be just as on top of the information as those farmers.

Foodborne bacteria and viruses do not make themselves known in the chicken's droppings. They are microscopic, and we know that manure contains a variety of detrimental organisms. From something as ubiquitous as *E. coli* to something a little rarer, such as *Salmonella enteritidis*, you can compost manure to remove as many of these organisms as possible.

The rule of thumb for organic production of vegetables and fruits is that raw manure be incorporated into the soil at least 120 days before the harvest of crops. That is specifically in reference to crops that come into contact with the soil surface. For those crops that do not come into the soil surface, then 90 days of time is sufficient. However, research has indicated that some particularly persistent organisms that are known to make people very sick have been able to survive beyond the 120 days.

The verdict

Poop. You should not let your chickens out into the garden or place raw manure into the garden when your plants are in the ground already. Inoculating the soil with fresh manure while plants are in the ground is a high-risk behavior and is not recommended. So, should you compost your manure? It certainly cannot hurt to use properly composted manure to help ensure that the level of safety of the food that you feed yourself and your family is high.

"Dunking your eggs in water should be part of washing your eggs."

☐ FACT ☐ POOP

TOPIC
HOME SWEET HOME NO. 8

ANSWER PROVIDED BY
Maurice Pitesky, DVM, MPVM, DACVPM
Veterinarian/Assistant Specialist in Cooperative Extension Poultry Health and Food Safety Epidemiology
School of Veterinary Medicine
University of California, Davis

The research

In my experience, the discussion about whether or not to wash your backyard eggs is about as controversial as religion and politics for many backyard poultry enthusiasts. It's just something you don't talk about in polite company. So, I'm going to duck that question and just answer the more specific version, which is whether or not dunking the eggs is part of washing them. The answer is, simply, "No."

The reason is that eggs, which are made of calcium and magnesium carbonate, are porous, containing between 6,000 and 10,000 pores that form an imperfect outer layer. (I would have hated to have been the graduate student who had to count.) The physiological reason for all those pores relates to the developing embryo. The eggshell itself is considered semipermeable, which means air and moisture can pass through its pores. In the case of the embryo, that means carbon dioxide and moisture.

The pores can also facilitate infection of the internal contents of the egg from bacteria on the surface of the shell and the external environment. Each unwashed egg can contain up to one million bacteria on its exterior, including common bacteria like *Pseudomonas*, *E. coli*, and *Streptococcus*. Most of these are nonpathogenic; however, dunking the eggs in water creates a situation in which the bacteria can move from the outside of the shell toward the inside. Though *Salmonella* is relatively rare in commercial table eggs sold in the United States (1 out of every 10,000 to 30,000 eggs), *Salmonella* food poisoning should still be considered a legitimate concern, especially in eggs not properly cooked (i.e., "runny eggs").

Although bacteria have to breach other internal structures such as the inner and outer membranes to get to the yolk, then grow and divide to produce enough to cause an infection if consumed, bacteria *can* move beyond these structures. Therefore, practices that can facilitate bacterial transfer across the shell—in this case, dunking them in water—are never a good idea and are counter to the original idea behind the action, which is presumably to clean them to reduce the chance of foodborne illness.

The verdict

Poop. As noted earlier, to wash or not to wash backyard eggs is a controversial topic. In the United States, table eggs have to be washed in order to get sold commercially. However, those eggs are not dunked. So, regardless of whether you wash or don't wash, don't dunk.

“I need to have a rooster if I want eggs from my hens.”

☐ FACT ☐ POOP

TOPIC

HOME SWEET HOME NO. 9

ANSWER PROVIDED BY

Brigid A. McCrea, PhD
Extension Specialist
4-H Youth Development Animal Programs
Alabama Cooperative Extension System
Auburn University

The research

Hens will lay eggs daily, or almost daily, regardless of whether a rooster is present. This goes back to basic biology. Females lay eggs on a regular or seasonal basis as determined by their species. For example, humans release eggs once a month. Chickens release them daily when day lengths exceed fourteen hours. All female birds respond to the longer days, and their gonads grow to prepare for egg laying. They also respond to shorter days of fall and stop producing eggs.

Given this fact, why do hens continue laying through the winter? Well, we have selected them for increased egg production. We have domesticated chickens and selected them for certain aspects of production for hundreds (technically thousands) of years. So, they may lay well through their first winter, but you can expect a slowdown or stop in egg production in the following winters—that is, unless you provide supplemental light for the hens in the fall and winter. If you are providing supplemental light, then they will continue to lay in those seasons.

The verdict

Poop. This is quite the common misconception. You do not need to have a rooster to get eggs from a hen. In short, it is a biology and seasonal thing, not a rooster thing.

"Chickens and turkeys get along great. There are no problems raising them together."

☐ FACT ☐ POOP

TOPIC

HOME SWEET HOME NO. 10

ANSWER PROVIDED BY

Brigid A. McCrea, PhD
Extension Specialist
4-H Youth Development Animal Programs
Alabama Cooperative Extension System
Auburn University

The research

Many small flocks owners believe that chickens and turkeys have been kept together traditionally. Such is not the case. Until turkeys were brought to Europe from North America, the two species did not intermingle. Certain parasites common in chickens, like that which causes blackhead disease, can have a detrimental effect on turkeys. Chickens can get an internal parasite that can carry a protozoal organism called *Histomonas meleagridis*. This parasite causes sickness and sometimes death in turkeys. Regardless of whether the turkey lives or dies, the infection is very uncomfortable for the bird.

This transfer can be prevented with good biosecurity measures. One of the guiding biosecurity principles is to avoid mixing avian species on a farm. Keeping a single species removes many opportunities for certain pathogens and parasites to cause harm by moving from a potential host species to an unintended target. Once a flock is infected, problem organisms can spread in many ways. The trouble is, you cannot see the organisms or how they try to find new hosts.

From a management perspective, keeping together two different species like chickens and turkeys is a logistical challenge. The feeds and feeding requirements for each differ, such as the height of feed and watering equipment, especially when they are young and growing. The effort it takes to keep turkeys out of chicken feed, and perhaps vice versa, is really not a detail most small flock owners will want to deal with.

The verdict

Mainly poop. If possible, do not raise turkeys and chickens together. There is potential to spread unwanted disease, and from a management perspective, the two species each have different requirements.

“Chickens quit laying at two-and-one-half years old.”

FACT POOP

TOPIC

HOME SWEET HOME NO. 11

ANSWER PROVIDED BY

Brigid A. McCrea, PhD
Extension Specialist
4-H Youth Development Animal Programs
Alabama Cooperative Extension System
Auburn University

The research

The majority of chickens sold to small flock holders in the United States are brown-shelled egg layers. The strains sold are used in egg production. These strains have been bred and developed for egg production over the last 60 years.

Certain strains of chickens are bred for consistent and uniform egg production. Eggs should be of high quality with regard to both the interior and exterior factors. When a hen begins laying for the very first time in her very first laying cycle, her eggs will be small. As her body continues to adjust and grow, she will lay larger eggs. In her first laying cycle, a hen typically lays eggs of USDA size Peewee, Small, Medium, and Large.

When a hen begins to reach the end of her laying cycle, she begins to run out of resources for consistent egg production. Her body needs a rest, and so she goes into a molt. Molt is when the hen ceases egg laying and replaces her feathers. As her egg production begins to taper off, the size, consistency, and quality of her eggs fall off over a couple of weeks before she is in full molt. In commercial egg production, these details are carefully monitored to determine when it is a good time to place the entire barn into a molt.

If a flock of commercial birds begins to taper off in egg production to the tune of approximately 80 to 85 percent, then it is time to molt the barn. This usually occurs when the flock is sixty to seventy weeks old. When the hens come back into lay in their second cycle, then the size of the eggs that are laid often increases to include Extra Large and Jumbo sizes. The profits are greater from the sale of larger eggs. The rate of egg laying and the number of weeks of egg laying decreases during the second cycle.

Usually, the shells are thinner on these larger second-cycle eggs. The hen only has a finite amount of shell that she can place on an egg, regardless of the size. So, as an egg gets larger, the shells are thinner. This is a consideration for the second-cycle flock. If the hens are laying eggs that have shells so thin that they break on their way to the market, then it may be time to molt the flock again. Several factors, including shell thickness, affect this decision.

The economics of egg production means that flocks are sold to a processor if they are not producing at a good enough rate to warrant continued housing and feeding. The market forces around certain times of the year mean that a flock could be molted again and then lay for a third cycle before they are taken to a processor. However, these birds will produce eggs for a shorter period and at a much reduced rate of lay.

The egg layer breeding companies, after many years of consultation with farmers and companies, have developed a bird that can nearly guarantee a profitable rate of lay for the first two cycles. This provides the companies, and their associated farmers, with a sort of guarantee that their investment of time, energy, and money into a flock will yield results for at least the first two cycles. The first two cycles is about two-and-one-half years of age after the chick hatches. After that point, a company and farmer need to closely examine market forces to determine if the flock will be retained or processed.

The verdict

Hens lay with the greatest efficiency in their first 2.5 years. They may, perhaps, continue laying for a few more years with increasing inefficiency, so if you got into keeping chickens for the eggs, then you will need an exit plan for when they stop laying completely. Are you OK with keeping a flock of nonproductive hens until they all pass, knowing that good biosecurity precludes you from obtaining a new flock until the old one is gone? Remember that chickens live into their teens. Although small flock owners tend to keep their flocks until they die, they do not constitute the majority of buyers of these particular egg-laying strains. With regard to noncommercial strains or heritage breeds of chickens, many of them are not assessed, or bred, for a high rate of egg production.

> **"Backyard poultry are less likely to produce eggs contaminated with *Salmonella* than commercial egg-laying poultry."**

Chicken poop. All poultry can carry germs such as *Salmonella*. Additionally, researchers at Pennsylvania State University conducted a large survey looking at the *Salmonella* prevalence in eggs from small chicken flocks purchased from farm stands across Pennsylvania. The survey found that 2 percent of the eggs tested were positive for *Salmonella*, and the contamination was primarily found inside the egg. The number of eggs from small flocks testing positive for *Salmonella* was much greater than the percentage of commercial eggs testing positive for *Salmonella* (less than 0.5 percent).

ANSWER PROVIDED BY Megin Nichols, DVM, MPH, DACVPM

Enteric Zoonoses Activity Lead Centers for Disease Control and Prevention

"Fermenting feed will cut my feed bill in half."

☐ FACT ☐ POOP

TOPIC

HOME SWEET HOME NO. 12

ANSWER PROVIDED BY

Nancy Buchanan Jefferson, PhD

Poultry Nutritionist

Kalmbach Feeds

The research

If fermentation is conducted correctly such that the pH of the finished product is properly reduced, there is evidence that chickens gain more weight (Chen et al., 2009, *Poultry Science*). However, even perfectly fermented feed does not reduce intake by half.

A study of young laying hens fed a diet of fermented feed versus dry feed showed only a 12 percent reduction in feed intake (on a dry matter basis) for the hens consuming the fermented feed (Engberg et al., 2009, *British Poultry Science*). Moreover, feed intake actually increased by 5 percent (on an air-dry basis) for broiler chickens consuming a fermented feed compared to a dry feed.

Despite evidence that proper feed fermentation alters the nutrient profile, most of the research points to small changes. A comparison of crude protein and energy availability showed no difference in available crude protein but a small increase (about 1.5 percent) in available energy for broilers consuming a properly fermented feed compared to a dry feed. In a laying feed, there was a small increase in crude protein (about 3 percent) but no change in energy for a properly fermented feed compared to a dry feed. And there was a decrease in the essential amino acids lysine (6 percent) and threonine (3 percent) in the fermented feed.

On a positive note, it appears that properly fermenting feeds does decrease the amount of phytate-bound phosphorus in a feed by nearly 30 percent. Releasing bound phosphorus is important for the chickens' bone development and growth and is traditionally accomplished by using a dietary enzyme.

The verdict

Mainly poop. There is evidence that feed fermentation has a positive effect on the intestinal microflora of a chicken. It appears that properly fermenting poultry feed benefits the gastrointestinal microflora most—rather than decreasing feed intake and your feed bill—by helping grow beneficial bacteria and reduce pH in the gut that can help mitigate the negative effects of pathogenic bacteria. However, changes to the nutrient profile of the feed and the digestibility of the feedstuff are relatively small.

"The egg's bloom is sacred and must never be breached."

☐ FACT ☐ POOP

TOPIC

HOME SWEET HOME NO. 13

ANSWER PROVIDED BY

Brigid A. McCrea, PhD
Extension Specialist
4-H Youth Development Animal Programs
Alabama Cooperative Extension System
Auburn University

The research

If you wish to hatch eggs, then one of the egg's strongest defenses against bacterial or fungal invasion is the bloom. So what is the bloom, or cuticle, of the egg? The bloom is the outermost layer of the egg and is put on last in the egg-formation process. The bloom consists of glycoproteins that partially block the many thousands of pores in the eggshell. So . . . surprise . . . if you did not already know that the eggshell is porous, then you do now! The main function of the bloom is to help lubricate the egg in the laying process. Being made mostly of water, the bloom quickly dries after oviposition, thereby blocking the pores.

The pores are all part of the eggshell design. By that, I mean all eggshells have pores. Air enters the pores and water exits the pores. If there is a baby chick inside, then carbon dioxide exits through the pores as well.

The bloom does break down on its own as a natural function of the incubation process. As the bloom breaks down, the pores open up a little more. This corresponds with the oxygen and carbon dioxide exchange needs of the developing embryo. The number of pores and the amount of bloom placed on the egg varies based on the bird species and the age of the hen laying the egg.

"If the outside of my eggs are clean, I can safely eat them raw or undercooked."

Chicken poop. Eating eggs that are raw or undercooked, even if clean on the outside, can put you at risk for *Salmonella infection. Salmonella* can live inside the egg even when it appears normal and is cleaned, which is why eggs should be cooked until both the yolk and white are firm with an internal temperature of at least 160°F (71°C). Eggs should not be kept at room temperature for more than two hours or one hour if the temperature is 90°F (32°C) or warmer.

ANSWER PROVIDED BY Megin Nichols, DVM, MPH, DACVPM
Enteric Zoonoses Activity Lead Centers for Disease Control and Prevention

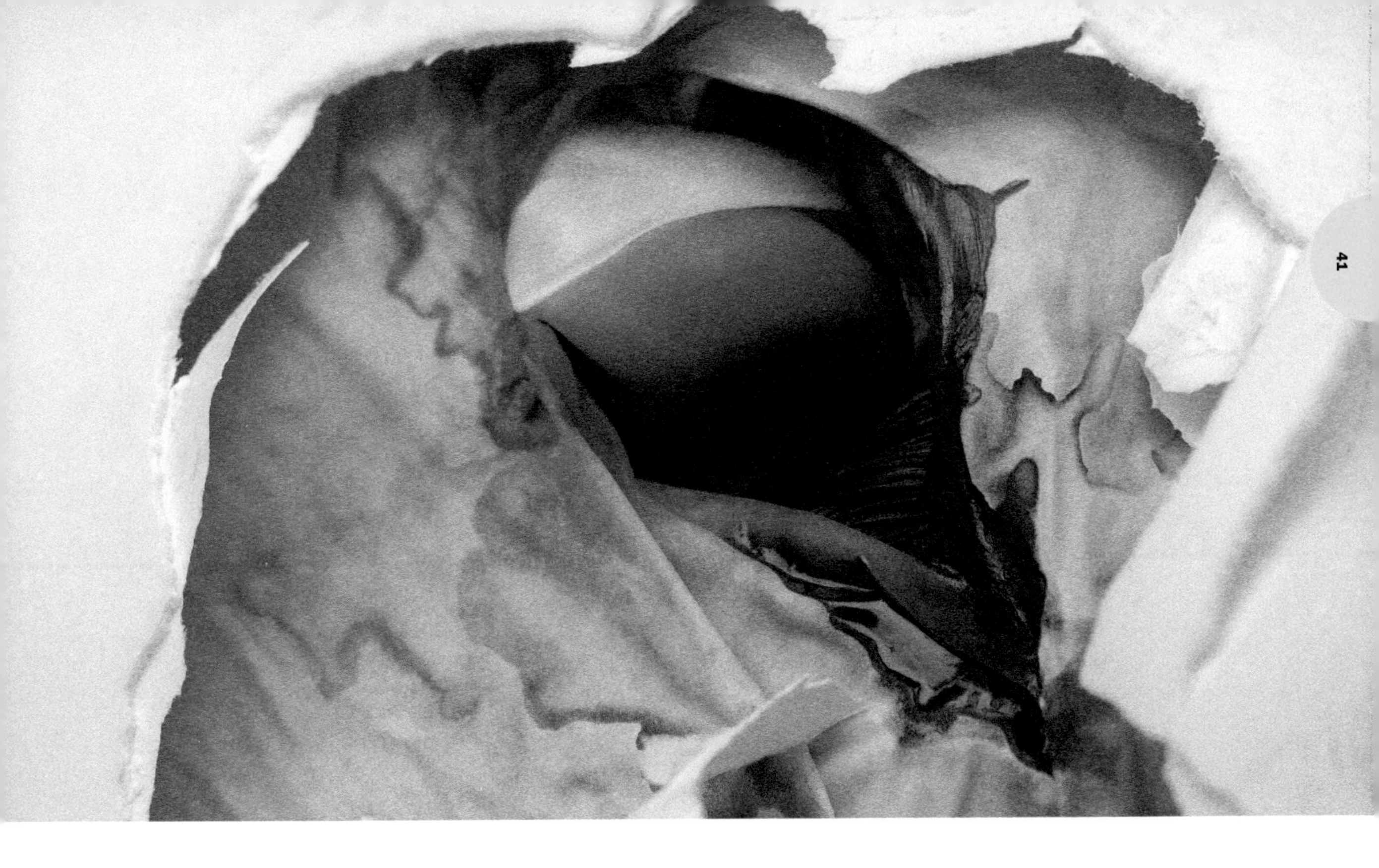

For table eggs, the bloom is not necessary. In the United States, eggs produced commercially are all washed at a temperature that is warmer than the warmest egg. This is a food safety control point to help keep consumers from becoming ill by eating unclean eggs (mostly because companies have no control over how the eggs are handled after they are purchased and mistakes by consumers have been made many times). So in the washing process, the bloom is removed. It used to be a common practice to spray the eggs with an extremely light coating of odorless, tasteless mineral oil. But this step is no longer administered in many cases. It was too costly of an operation, and egg companies are very efficient entities.

The verdict

Mainly poop. But to some degree the statement is true. So, where does this leave you, the backyard flock owner? Well, since your chickens are not likely separated from their feces in the same way that a commercial flock is, then it is likely that your management challenges are quite different. You are not likely to know what organisms are affecting your flock on a weekly basis, so it is best to wash your eggs daily. We refrigerate eggs in our country, so washing the eggs and removing the bloom is not a worry.

The only reason to not wash an egg is if it is perfectly clean and you intend to use it for hatching. Otherwise, wash away that bloom and any bacteria that may be hanging out on the outside of the shell. Then, stick those eggs in the fridge, and you can count on a high-quality egg waiting for your next culinary adventure!

TEMPERATURE AT WHICH THE EMBRYO INSIDE THE EGG STARTS TO GROW: 86°F (30°C)
WEEKS IT TAKES FOR A FEMALE CHICK TO MATURE AND LAY EGGS: 18–24
HOURS IT TAKES FOR A HEN TO PRODUCE AND LAY AN EGG: ABOUT 26
DAYS FOR A BABY CHICK TO HATCH
21

"You can de-crow a rooster just like you de-bark a dog so they won't be as noisy."

☐ FACT ☐ POOP

TOPIC

HOME SWEET HOME NO. 14

ANSWER PROVIDED BY

Brigid A. McCrea, PhD
Extension Specialist
4-H Youth Development Animal Programs
Alabama Cooperative Extension System
Auburn University

The research

The anatomy of a chicken is different from that of mammals. If you have a troublesome dog that barks, you can make an appointment for a quick surgery to reduce the severity of the bark. That does not take care of the frequency of barking; rather, that will take some training. Since the situation works for dogs, many small flock owners have wondered about de-crowing a rooster.

The vocal cords for a human or dog are located in the throat. The chicken, and birds in general, have a syrinx, which they use for their song. A rooster's crow is just a chicken version of a song or territorial call. The syrinx is located at the bifurcation of the bronchi deep within the respiratory system—within the body cavity of the bird, not the neck. That makes it inaccessible to the same surgical procedure of de-barking in dogs.

The verdict

Poop. There have been alternative strategies, not all of which are effective. Some researchers have tried chemical castration of chickens with mixed results. The results are only temporary, and it is not really being pursued as a solution at this time (Kilger, 1950). Alternatively, some people have caponized their roosters. This is an attempt, most often, more at reducing aggression rather than reducing crowing. Caponizing is when the testicles are removed from a male chicken. This is akin to neutering a dog or castrating a lamb, pig, or any other male farm animal. Caponizing usually increases the gain of the males, which is why this is an option for male chickens.

Sometimes, caponizing works in reducing aggression, but what most people find off-putting is again that you need to enter the body cavity because that is where the testicles are located. This is unlike mammalian testicles, which are outside the main body cavity, making castration easier. Also, this is most often done without the aid of general anesthesia (Mast et al., 1981). There is a crowing-reduction collar out there that affixes tightly to the neck of a rooster. There is little to no research as to the efficacy of this device. So, the myth that you can de-crow a chicken in the same manner as you de-bark a dog is debunked. But not to worry: At least your rooster will wake you up at first light in the morning!

2

Safe and Sound: Protecting Your Flock

"Putting hot sauce in your chickens' water will keep them warmer in the winter."

☐ FACT ☐ POOP

TOPIC
SAFE AND SOUND NO. 1

ANSWER PROVIDED BY
Maurice Pitesky, DVM, MPVM, DACVPM
Veterinarian/Assistant Specialist in Cooperative Extension Poultry Health and Food Safety Epidemiology
School of Veterinary Medicine
University of California, Davis

The research

The compound capsaicin and its relative compounds produce a burning sensation (i.e., the spiciness we experience when we eat it). Interestingly, birds do not have taste receptors for capsaicin. In addition, even if they did have taste receptors for capsaicin, eating spicy foods would not raise their body temperature in an effective way to help them maintain a warmer body temperature continuously.

So, how do birds stay warm? Feathers, for starters. Aside from being a way we can tell apart different poultry species, feathers serve to help birds retain heat and stay dry by trapping an insulated layer of air against the skin. Down feathers compose the layer closest to the body and provide the most insulation from the cold.

In addition, birds, like mammals, are warm-blooded, meaning they maintain a constant body temperature independent of the outside environment. The body and all its enzymes typically work better in a very narrow temperature range. Therefore, expending energy to maintain a very narrow temperature range allows birds and mammals to operate at a high level of efficiency. This requires energy, which birds—and humans—get from food. In contrast, anyone who has ever seen a cold-blooded reptile on a rock trying to warm up understands how slowly it moves. The advantage for the reptile is that it doesn't have to waste the energy from food on maintaining an independent temperature.

The verdict

Poop. Big picture, feathers and the physiological ability to maintain an independent body temperature keep chickens warm, not hot sauce in their water.

“Apple cider vinegar in chickens’ drinking water makes for stronger eggshells.”

☐ FACT ☐ POOP

TOPIC

SAFE AND SOUND NO. 2

ANSWER PROVIDED BY

Maurice Pitesky, DVM, MPVM, DACVPM
Veterinarian/Assistant Specialist in Cooperative Extension Poultry Health and Food Safety Epidemiology
School of Veterinary Medicine
University of California, Davis

The research

Eggshell quality is a significant problem in commercial eggs throughout the production, processing, and distribution pathways. A study in 1977 by Roland in the journal *Poultry Science* determined that 7.77 percent of eggs are lost in the hen house due to poor shell quality. Additional losses, about 6 percent, occur between collection and consumer.

Factors that contribute to shell damage include the birds’ genetics, health, nutritional status, and the environment and management in which they are raised, as well as the mechanics of collecting, washing, packing, transporting, and storing the eggs.

For example with respect to nutrition, assuming a hen eats properly, dietary levels of calcium, phosphorus, vitamin D3, manganese, and copper are critical for eggshell quality. For adequate calcium, the dietary level, source, particle size, and feeding time can all impact eggshell quality. (Interestingly, feed that delivers dietary calcium at night, during the physiological time of shell formation, also positively affects this process; this is one advantage of feeding the birds larger oyster shells. Due to their size, it takes longer to metabolize and allows for more calcium absorption.)

These same factors affect backyard chickens. You may have noticed that as your hens age, their eggs generally get bigger. Even so, a small egg and a jumbo egg receive the same calcium deposits. Therefore, the jumbo egg has a thinner shell, which is consequently more likely to crack. To provide for optimal egg-production conditions, use a reputable commercial feed and provide a quality environment for the hens.

The verdict

Poop. Eggshells are made of two ingredients: approximately 98 percent calcium carbonate and 2 percent magnesium carbonate. Apple cider vinegar has zero calcium and zero magnesium per 1 tablespoon (15 ml) serving. In short, it does not contribute to shell strength.

"Provide quality grit, and there is no need to worm your chickens."

☐ FACT ☐ POOP

TOPIC

SAFE AND SOUND NO. 3

ANSWER PROVIDED BY

Brigid A. McCrea, PhD
Extension Specialist
4-H Youth Development Animal Programs
Alabama Cooperative Extension System
Auburn University

The research

Grit are small stones consumed by chickens and a few other bird species. They are found in the gizzard and, along with the gizzard's strong lining, help to grind up seeds and grains that the birds eat. This process increases the animals' surface area inside the gizzard where grains may be crushed and broken down before passing into the small intestine for absorption. Grit eventually wears down, passing through the intestine and out in the feces; therefore, birds need to consume stones regularly to keep enough in their gizzard to assist with the grinding action.

Grit, when it finally reaches the small intestine, is very small. It does not push out internal parasites. In the 1950 article "The use of grit and its effect upon *Ascarid* infections" in *Poultry Science* by Bernard Riedel, research proved that chickens fed grit did not have any more or any fewer roundworms than chickens fed a diet without grit.

Since worms are not removed physically by things consumed by the bird, such as grit, scientists have found alternatives. Piperazine is the active ingredient in the over-the-counter product Wazine, and it helps to remove internal parasites such as worms. It does not kill the worms; rather, it puts them to sleep and they will hopefully let go of the gut lining. Once free in the lumen of the intestine, normal peristaltic action pushes the worms out and into the feces. From there, the worms will hopefully die after the feces are cleaned up. In a worst-case scenario, a chicken may turn around and consume a worm if they see it moving in the feces. Chickens are attracted by movement and could reinfect themselves via the fecal-oral route.

Other worming products do exist but require a veterinarian's prescription and oversight since their use is considered an extra-label use. The legalities of administering drugs without veterinarian oversight has been of great concern of late.

The verdict

Poop. Grit will not remove internal parasites from the gastrointestinal tract of a chicken.

“Mites live happily on the vents and the warm undersides of chickens.”

☐ FACT ☐ POOP

TOPIC

SAFE AND SOUND NO. 4

ANSWER PROVIDED BY

Brigid A. McCrea, PhD
Extension Specialist
4-H Youth Development Animal Programs
Alabama Cooperative Extension System
Auburn University

The research

There are three types of mites commonly associated with chickens: scaly leg mites, northern fowl mites, and red chicken mites. For the above statement to be true, mites would have to exist only around the vent and warm undersides. This is not so. Scaly leg mites are found on the scaly, keratinized regions of the body like the legs and the beak. Because they are microscopic, they are not visible with the naked eye.

Northern fowl mites are found around the vent and warm undersides. Honestly, they can show up on any part of the body, but they prefer feathered areas because that's where they lay their groups of eggs. They also chew on the skin and parts of the feathers, as well as dine on scabs and bird blood. They tend to hang out near the skin but can be found on any part of a feather.

Red chicken mites tend to live in the birds' environment. Most people cannot even begin to fathom how many red mites exist in the environment of their coops. Why? Because these mites are good at hiding in the cracks, nooks, and crannies of the wood of most chicken coops. They also hide in groups under clumps of caked litter. Most often they leave their environment at night to feed on the birds. That is not to say that you cannot find them on birds during the day, but your chances of seeing them is greater at night. They turn red, but only after a blood meal, so they can look like northern fowl mites at other times.

The verdict

Mainly poop. Some mites do live on the vents and warm undersides of chickens, namely northern fowl mites. But the two other types of mites that afflict chickens—scaly leg mites and red chicken mites—don't. Instead, the former inhabit keratinized regions of the body, like the beak and scale, and the latter prefer the environment of the chicken coop itself.

"I need to have a rooster to protect my flock."

☐ FACT ☐ POOP

TOPIC

SAFE AND SOUND NO. 5

ANSWER PROVIDED BY

Brigid A. McCrea, PhD
Extension Specialist
4-H Youth Development Animal Programs
Alabama Cooperative Extension System
Auburn University

The research

In the harem of wild chickens (a.k.a. red jungle fowl), the male takes on three roles, according to a 1967 study from N.E. Collias and E.C. Collias published in *The Condor*. The rooster defends and maintains a territory, he protects the hens and young birds, and he sires the next generation of young birds.

In most backyard chicken coops, the size and style of the coop that the owner provides negate the first two points. If the coop protects the flock from both ground and aerial predators, then protection by the rooster is not needed. You, as property and flock owner, also protect and defend the space in which the hens live. You may have a secure yard with a fence or perhaps you only let your hens out when you can be outside with them.

For the final point of a rooster providing genetic material, it is not required if you are simply keeping a backyard flock for egg production. See page 31 for more information about that.

In the war years, when protein sources were being rationed and sent to the troops, keeping a flock of chickens at home was a great plan. It meant that you could have a steady source of protein in your diet and the diet of your family members. It was recommended and promoted by states that the male chickens be removed, or consumed, in order to help families keep more of their eggs. Why could keeping a rooster affect the eggs? Refrigeration was not widely used, and if an egg was exposed to high temperatures before being collected, like the temperatures you would find in the summer months, then you could begin to see embryo development. A developing embryo is not what consumers want to see, and that can lead to the egg being thrown out and wasted. A developing embryo also changes the nature of the eggs contents after short period of development (i.e., as few as one or two days), making them less suitable for cooking or baking.

The verdict

Poop. Keeping a flock of backyard chickens does not require the inclusion of a rooster if you have a well-designed coop that protects the hens and seeing a developing embryo inside a table egg is off-putting to consumers.

"Chicken coops need heat."

 FACT POOP

TOPIC

SAFE AND SOUND NO. 6

ANSWER PROVIDED BY

Brigid A. McCrea, PhD

Extension Specialist

4-H Youth Development Animal Programs

Alabama Cooperative Extension System

Auburn University

The research

Chickens are birds of the bamboo forest. They are designed for fairly warm climates. They have strategies for both keeping warm and expelling heat. However, humans have greatly influenced the appearance of the chicken. Some breeds, such as the Chantecler, have been selected for their adaptability to colder climates. So what does a cold coop really mean to a chicken?

Chickens lose heat through their combs. Males have larger combs and wattles and in the winter, are more prone to frostbite on both. Hens have smaller combs and wattles and therefore less surface area. Not all chickens put their heads under their wings at night. Even if they do, breeds with large combs may not be able to completely tuck the comb under their wing. Any exposed flesh is prone to frostbite at night, when temperatures are at their coldest.

Toes can also get frostbite. Chickens like to pull their toes up under the feathers on their breast. The shape of the roost in the coop can either help or hinder this process. A wider, flatter roost will allow birds to tuck up their toes and any unfeathered areas on their legs.

Egg laying can be forfeited for many reasons, including suboptimal environmental conditions.

So, what are the critical temperatures for frostbite? Anything below freezing, which is 32°F (0°C) or lower. What about temperatures just above freezing? When the temperature of the coop drops below 55°F (13°C), hens start to slow down the egg-making process. Prolonged low temperatures may start to yield smaller eggs or fewer eggs. Temperatures below freezing for prolonged periods reduce this rate of lay much faster. When temperatures get below 0°F (-18°C), your flock may stop laying eggs altogether. A hen will stop producing eggs in favor of staying alive.

The verdict

Fact and Poop. You may find providing heat to be out of the question on your farm. Luckily you have other options in the form of good insulation and winterization of the coop. How do you insulate? It can be as easy as stacking hay bales around the walls or putting up insulation on the roof and walls. Chickens eat insulation so be sure to cover it up with plywood. Winterization can include wrapping draffy portions of your coop with in thick plastic.

If you rely on the eggs from your flock, I would recommend providing supplemental heat when temperatures outside approach freezing. How do you know the coop temperature? Keep a min/max thermometer in the coop and record the lowest temperatures on a clipboard. Alongside temperature, record the number of eggs you get daily to determine any noticeable trends within your flock.

For warmth, you could try a heat lamp, though I dislike them generally because they provide not only heat but light as well. Over-lighting your hens can also halt egg production because the hen's reproductive tract gets tired. A heat lamp on at night makes the hen thinks it is still daytime physiologically, so her body continues to respond to the long days. You might not see any big changes in young hens, but older hens may quit laying sooner and go into a molt at an inopportune time of year.

One other reason I dislike heat lamps is because they can start fires from being knocked over or broken. I have seen rodents chew through rope, causing the lamp to drop into the shavings (so use a chain to hang heat lamps). I have also seen lamps shatter when chickens shake cold water from their beaks and it lands on the bulb.

I prefer safer alternatives like the Sweeter Heater. Hang it over the roosts and plug it into a Thermo Cube so it turns on and runs only when the cold is of concern to your flock or plug into a light timer so that you only pay to run it when the coop is coldest at night.

What it means

If you expect eggs from your hens, be prepared to provide a form of heat. Older hens, sick chickens hiding an illness, or birds with parasites go into winter already at a disadvantage and may not make it through the first cold snap. To prevent a negative experience for your birds, you may wish to put a heat source in the coop, ideally on a timer or on a Thermo Cube that automatically turns on when temperatures dip to their lowest.

"You should protect baby chicks by bringing them into your house."

ANSWER PROVIDED BY

Megin Nichols, DVM, MPH, DACVPM,
Enteric Zoonoses Activity Lead Centers for Disease Control and Prevention

Chicken poop. Baby chicks, just like adult poultry of any kind, should not be brought into the house. This is because of the many ways they can pass on germs such as *Salmonella* and *Campylobacter* to people. Between 1990 and 2014, fifty-three live poultry–associated *Salmonella* outbreaks were recorded in the United States, leading to 2,630 illnesses, 387 hospitalizations, and five deaths. Nearly half of all *Salmonella*-infected patients reported keeping poultry inside the house. Bringing live poultry, such as chicks and ducklings, into the house puts everyone in the household at risk for infection and illness. A better method is to keep poultry in their own outdoor chicken coop or brooder.

"Every hen needs her own nest box."

☐ FACT ☐ POOP

TOPIC

SAFE AND SOUND NO. 7

ANSWER PROVIDED BY

Brigid A. McCrea, PhD
Extension Specialist
4-H Youth Development Animal Programs
Alabama Cooperative Extension System
Auburn University

The research

Many factors influence which nest box hens in the coop use. Colony nest boxes that are a little larger allow many hens to use a nest box but require different management methods to prevent broodiness, broken eggs, or even egg-eating behavior.

You can reduce the frequency of egg losses or decreases in shell quality when you have no more than four hens to a nest box. That does not mean that the flock will intuit your intention in providing a certain number of nest boxes. Your role as flock owner is to prevent mishaps and make management corrections if the hens choose to all lay in one box or, worse, not use the nest boxes at all.

The verdict

The answer to this is 100 percent poop. Every hen does not need her own nest box. Hens are quite capable of using a communal nest box and quite often do under certain circumstances.

What it means

There are hens who lay in nest boxes, and then there are hens that lay on the floor. The reasons behind this preference have been explored by researchers in various trials over the years. It is preferable to have hens that lay in next boxes because the eggs are easier to retrieve, cleaner, and less likely to be broken by being stepped on in the coop. But why do some hens persist in laying on the floor despite all the efforts of flock owners to encourage them to lay in the next box? Researchers in Switzerland (Zuchen et al., 2008) designed a study looking at hen laying behavior and nest selection preference during their first twenty eggs. The behavior of hens that laid on the floor was more unsettled than that of hens that laid in nest boxes. However, once the hens chose to lay on the floor or in nest boxes, they exhibited fidelity in that site. That means that if a flock owner with newly laying hens finds a hen laying in the litter, then they should consider encouraging their floor layers to lay in the nest box right away.

Why do some hens lay more often in one nest box and less often in others? This is an observation made by many backyard flock owners. It has been observed in some studies that hens exhibit a preference based on the position of the nest box in the coop. Whether it is the relative darkness of the nest box or its position relative to the door leading to the run, hens spend a fair amount of time exploring their options before choosing where they lay (Appleby et al., 1984).

Some have considered that perhaps the social status of the hen might play a role in her choice of nest boxes. The social status of a hen does not play a role but can affect the behavior of the hen just before she lays. If a hen is lower in the peck order ranking within the flock, then she is more likely to make more frequent visits to the nest box before she lays her egg (Ringgenberg et al., 2015).

Overall, it is well recognized that the ratio of one nest for every four hens is sufficient for egg production if you are using individual nests. If using individual nests, smaller-bodied hens, like that of the Mediterranean breeds (e.g., Leghorn or Ancona) do fine in the small-size nests found within most common chicken coops. Those nests tend to be 12 inches (30.5 cm) in length, width, and depth. However, an extra 2 inches (5 cm) in depth is recommended for commercial egg production. For the larger, dual-purpose hens (e.g., Orpington or Plymouth Rock), another 2 inches (5 cm) in the width is more appropriate (Bell and North, 2002). Even larger breeds will need even more space.

So, that means if you have a Leghorn, then 12 × 12 ×12 inches (30.5 × 30.5 × 30.5 cm) works, whereas for a commercial brown egg–laying strain (e.g., production reds or Red Stars), then a 12 × 12 × 14 inches (30.5 × 30.5 × 35.5 cm) is preferred. Then again, if you have a more ornamental or larger-bodied breed (e.g., Cochin or Brahma), then you should consider a nest box that is even larger to meet their specific body size or tail length.

There are three other types of nests that people do not often explore. They are community or colony nests, roll-away nests, and trap nests. Each has its own specific purpose, but for most backyard chicken owners, single nests work fine. If single nests are not working out well, then consider taking out the dividers to create something akin to a community nest. A community nest is 2 feet (61 cm) wide and 8 feet (244 cm) long with an opening at each end to allow hens to enter and leave. A community nest will accommodate up to sixty hens (Bell and North, 2002).

It is also recommended with a community nesting system, as with all nesting systems, that it be closed at night to prevent hens from sleeping in, and thereby soiling, the nest boxes. Allowing your hens, or even your pullets, to sleep in the nest boxes is a bad management practice. Automatic doors set on a timer help flock owners to shut doors without the requirement that they be home to perform the task, and some systems can be adapted to the nest box entrance.

"Round roosts are better than flat roosts."

 FACT ☐ POOP

TOPIC

SAFE AND SOUND NO. 8

ANSWER PROVIDED BY

Brigid A. McCrea, PhD
Extension Specialist
4-H Youth Development Animal Programs
Alabama Cooperative Extension System
Auburn University

The research

Many small flock owners and coop-building companies pay almost no heed to the size, shape, and material of a chicken roost; but poultry scientists pay attention to every detail of chicken keeping, even the size and shape preference of roosts by hens housed in enriched conditions. These details often result in measurable differences in dollars and cents.

For example, as it turns out, perching is a learned behavior; chickens do not hatch out of their eggs already knowing how to perch. Yet, they still have a preference for roosting perch size. Previous research has shown that chickens with no prior experience roosting preferred round roosts to square, peaked, or flat roosts. They also preferred a larger diameter roost of almost 2 inches (5 cm) to those of two smaller diameters (Muiruri et al., 1990, *Applied Animal Behaviour Science*).

A similar study (Pickel et al., 2010, *Applied Animal Behaviour Science*) examined behaviors by laying hens. The researchers looked at perch diameter and material—including wood, steel, and rubber-covered perches—then cross-referenced the preferences with balance behaviors, resting periods, standing, and preening. As the diameter of the perch increased, regardless of its material, the number of balance behaviors decreased. In other words, it was easier for the chickens to keep their balance with larger perches. On steel perches, chickens rested with their heads tucked under their wings for longer periods regardless of the perch diameter.

Yet more research has found mixed results with regard to perch shape and bumblefoot incidence in chickens. Oddly shaped perches—mushroom or T-shaped, for example—tend toward more occurrences of bumblefoot than round or rectangular perches. Keel bone deformations are fewer in chickens kept on flatter perches compared to those that are circular.

Grip strength is also important to perch type. Something with a slightly rough surface helps chickens get a good grip. That means wood is preferred

over metal or plastic. Hardwood is more durable than softwood, which is prone to wear. The downside is that wood provides a great surface for red mites to hide.

Finally, poultry protect their toes from frostbite during roosting by lifting up their breast feathers and setting them down over their toes on the roost. A roost small in diameter may not allow the bird to entirely cover its toes with its feathers.

The verdict

Fact. The size, shape, and material of a roost matter—and chickens do have preferences, despite this behavior being learned rather than innate. They seem to want round roosts, larger than 2 inches (5 cm) in diameter. Larger perches with only a slightly rough surface allowed them to keep their balance better and get a stronger grip. In addition, a poorly designed roost can have long-term effects on the overall health of the hen. Many hours are spent daily resting the keel bone on the roost, so deformities are possible with a bad design.

"Placing herbs in a nest box repels and kills mites, lice, and other parasites."

☐ FACT ☐ POOP

TOPIC

SAFE AND SOUND NO. 9

ANSWER PROVIDED BY

Brigid A. McCrea, PhD

Extension Specialist

4-H Youth Development Animal Programs

Alabama Cooperative Extension System

Auburn University

The research

Avian scientists over the years have examined many aspects of birds. From anatomy and physiology to behavior and breeding, much information has been gathered on several, but not all, avian species. Interestingly, examinations of nests, their shapes, and materials have yielded interesting information.

Some bird species incorporate green material periodically into the nest after the eggs have been laid and the chicks hatched. Why? Some believe that it is to improve the health of the chicks in instances when external parasites have become problematic. Many of the avian species included in this research have altricial young. That means the young are very small, poorly feathered, and fully dependent on their parents for food and protection. If they get external parasites, they are not easily able to dislodge them. The parents or other neighboring birds are often the source of the external parasites.

The parents, in some circumstances, brought specific herbs into the nest. Some of those herbs are known to deter insects such as external parasites. The length of their efficacy remains unknown in some cases. It may be that the herbs' effectiveness only remains as long as the herb material remains fresh.

Chickens, contrastingly, have precocial young. That means that within hours of hatching, young are capable of leaving the nest and identifying food and water and are capable of some modicum of thermoregulation. That means that the hen provides food and warmth while helping the chicks familiarize themselves to the food and water sources in the area. The nest can be abandoned if it is overrun with external parasites. The home area of a precocial chicken chick is much larger than that of the altricial avian hatchling in a nest. It could be that the repellent nature of certain herbs in the nest is effective simply because of the small confined nature of the nest.

The verdict

Mainly poop. Many small flock owners have attempted to apply this knowledge to their own flocks. In some cases, small flock owners have placed fresh or dried herbs in their coops and nests. However, the effectiveness of this practice remains to be researched. There are at least three different types of external parasites commonly found in and on backyard chickens and their coops. This does not include the sundry less common external parasites. There is just not a lot of information on the effectiveness of herbal blends on their ability to keep the parasites off chickens.

In some cases, in countries that tend to free-range their flocks, indigenous methods of parasite control have included the use of herbs. It is through a rich oral history that many of these herbs, herbal parts, or blends of herbs have been shared within the communities. Hopefully, in those cases, as well as in our own in the United States, we can continue to share information on the control of destructive external parasites in our flocks. If that oral history happens to include information on herbs for this purpose, then more power to the poultry!

"You can cure an egg eater by drilling a hole in an egg and filling it with hot pepper sauce."

☐ FACT ☐ POOP

TOPIC

SAFE AND SOUND NO. 10

ANSWER PROVIDED BY

Brigid A. McCrea, PhD

Extension Specialist

4-H Youth Development Animal Programs

Alabama Cooperative Extension System

Auburn University

The research

One of the worst things that a small flock owner faces is an egg eater. Curing an egg eater of its obsession often means that it must be sold. However, circulating all over the Internet are rumors and myths about ways to cure an egg eater. Most often is a story about feeding the chicken a hollowed-out egg filled with something hot like cayenne pepper, blended jalapeños, or a hot sauce of some type. The egg is then returned to the coop, and the egg eater is baited in for the surprise egg.

Evolutionarily speaking, this is a design for reptiles, avians, and mammals to avoid foods that may be noxious. Pain sensation in the mouth can come from heat, mechanical, or chemical stimuli. These stimuli tell the animal that if they don't stop what they are eating at that moment, then tissue damage may be next. This allows the animal to make changes in its behavior in order to avoid the offending food again in the future.

The trouble with this practice is that chickens only have twenty-four taste buds located in the mouth. We have many more taste buds than they do, with about 9,000 in our mouth (North and Bell, 2002). Chickens, or rather birds, do not react the same way as we do to spices.

A bird's actual level of sensitivity to the heat of chili pepper is due to its species. Within the chili is a chemical compound called capsaicin. Chickens are far less sensitive to this chemical compound than a human. A dollop of sriracha sauce that may make you break out in a sweat may not even register on the heat scale for a chicken. It is thought that avians, in general, evolved this insensitivity to capsaicin as an adaptation for seed dispersal. This helps the plant, of course, in that the chicken is not harmed, or deterred, by the chemical constituents that are meant to drive away species that are not going to help the plant spread seeds (Jordt and Julius, 2002).

In fact, since many options for feeding antibiotics to chickens have been removed, research has been done to determine if capsaicin might be a substitute. Research in the 1990s found that if fed for a certain period as an ingredient incorporated into the diet, the birds did better after being challenged with a dose of *Salmonella* (Tellez et al., 1993).

The verdict

Poop. Go ahead and drill a hole in an egg and give your chickens some hot sauce. It is not likely to have the same effect as if you or I were to take a bite. A better use of your time is to prevent egg-eating behavior in the first place by picking up your eggs regularly and feeding a balanced diet to keep those eggshells strong.

"Pendulous crop and impaction are never a problem for free-ranging birds."

☐ FACT ☐ POOP

TOPIC

SAFE AND SOUND NO. 11

ANSWER PROVIDED BY

Brigid A. McCrea, PhD
Extension Specialist
4-H Youth Development Animal Programs
Alabama Cooperative Extension System
Auburn University

The research

Pendulous crop and crop impaction are both problems that have been alleviated by current management methods in the commercial industry. That is not to say that the problem has completely disappeared. It still exists but appears only when there has been a management failure of one kind or another.

So, what is pendulous crop? It occurs when the crop is overextended and does not empty itself regularly. This overextension means that the bottom part of the crop is lower than the exit point to the posterior esophagus. Once a bird has pendulous crop, then it is often a death sentence, as it can never regularly empty its crop. The bird eventually dies of starvation or malnutrition.

How does pendulous crop start? Usually, impaction is the cause. An impacted crop is one that is full of something that is either too large or too long to easily be emptied into the rest of the digestive tract. This impaction stretches the walls of the crop. Water may be drunk by the bird, and it occasionally does drain out of the crop, so you do see birds drinking occasionally.

Birds have a high metabolism, so even a day or so of blockage to the digestive tract can rapidly incapacitate a bird. Their crops will be full, but the bird is wasting away. You may begin to detect signs of fermentation if you open the beak and smell the odor from the crop. Crop impaction may be fixed by a veterinarian's surgical procedure. Many people try to turn a bird upside down and massage the blockage out of the crop. This can lead to some level of aspiration of the contents into the respiratory system. If this should occur, now a respiratory infection is likely to be next on the agenda of things to deal with.

In commercial poultry, impaction tends to occur if the birds are raised on the floor and they consume copious amounts of litter or bedding. Sometimes, it can come from eating too many insects rather than their regular diet.

Feathers can be another source of impaction. If fed an imbalanced diet, then chickens can resort to eating feathers off one another, or molted feathers, in order to attempt to meet their protein needs. This can lead to a ball of feathers blocking the crop much in the same way as grass impaction.

The verdict

Poop. When birds are put out to free range, impaction and potentially pendulous crop can result from eating grass that is too big or too indigestible. Long grass blades get wound up within the crop and never proceed down the rest of the digestive tract. This long grass then catches all the food that the bird normally eats and prevents it from proceeding down the digestive tract.

So, if you are free-ranging your flock, then make sure you keep the grass cut low. Tender grass shoots or legumes are much easier for the birds to digest. Keep your lawns and weeds trimmed to avoid a vet bill due to grass impaction.

Alektorophobia is the fear of chickens.
Chickens digest their food in 2 hours—from pellet to poop.

Chickens don’t have a bladder.
Wait . . . we don’t?

"When a chicken pecks at its coop mate, they are just saying hello."

☐ FACT ☐ POOP

TOPIC

SAFE AND SOUND NO. 12

ANSWER PROVIDED BY

Brigid A. McCrea, PhD
Extension Specialist
4-H Youth Development Animal Programs
Alabama Cooperative Extension System
Auburn University

The research

Chickens have no hands, no fingers, and therefore no way to give a handshake. And handshakes are really a human invention from years past. So how do chickens greet one another? Chickens are really not very interested in greeting one another. Rather, when meeting another chicken of the same sex, they busy themselves in seeking the newcomer's place in the peck order.

A chicken's rank in the peck order determines many things. It determines priority access to food and water, and it can also play a role in the frequency of mating if your flock has a rooster. Hens have a separate hierarchy from that of roosters. There is always a dominant hen, and she has the right to peck all hens that rank below her. The next in the hierarchy cannot peck the one above her but can peck all that rank below her. This continues down the hierarchy in the coop until you reach the bottom hen. She often is the most nervous in the coop and, if a rooster is present, may be willing to crouch for him with greater frequency.

That means that the lower-ranking hens may be missing a few more feathers on their backs or the back of their necks. That is not to say that roosters do not prefer certain hens, which may lead these hens also to be missing feathers in the same areas.

The verdict

Poop. If a hen is pecking a fellow member of the flock, it is not to say hello. It is usually a reminder of the hen's ranking within the flock. If it is a new member to the flock, then the chickens are likely reasserting their dominance in the flock. Removing or introducing a new chicken to the flock is a stressful time, as peck order rankings are up for grabs.

Another reason why a chicken may peck another chicken is due to nutritional deficiencies. If fed an imbalanced diet, chickens may start feather pecking, which can lead to cannibalism. If a diet is too high in energy and too low in protein, then chickens may consume their own feathers, molted feathers, or a few feathers off a lower-ranking hen. So, pecking one another has the possibility of either being a peck order or nutritional imbalance issue.

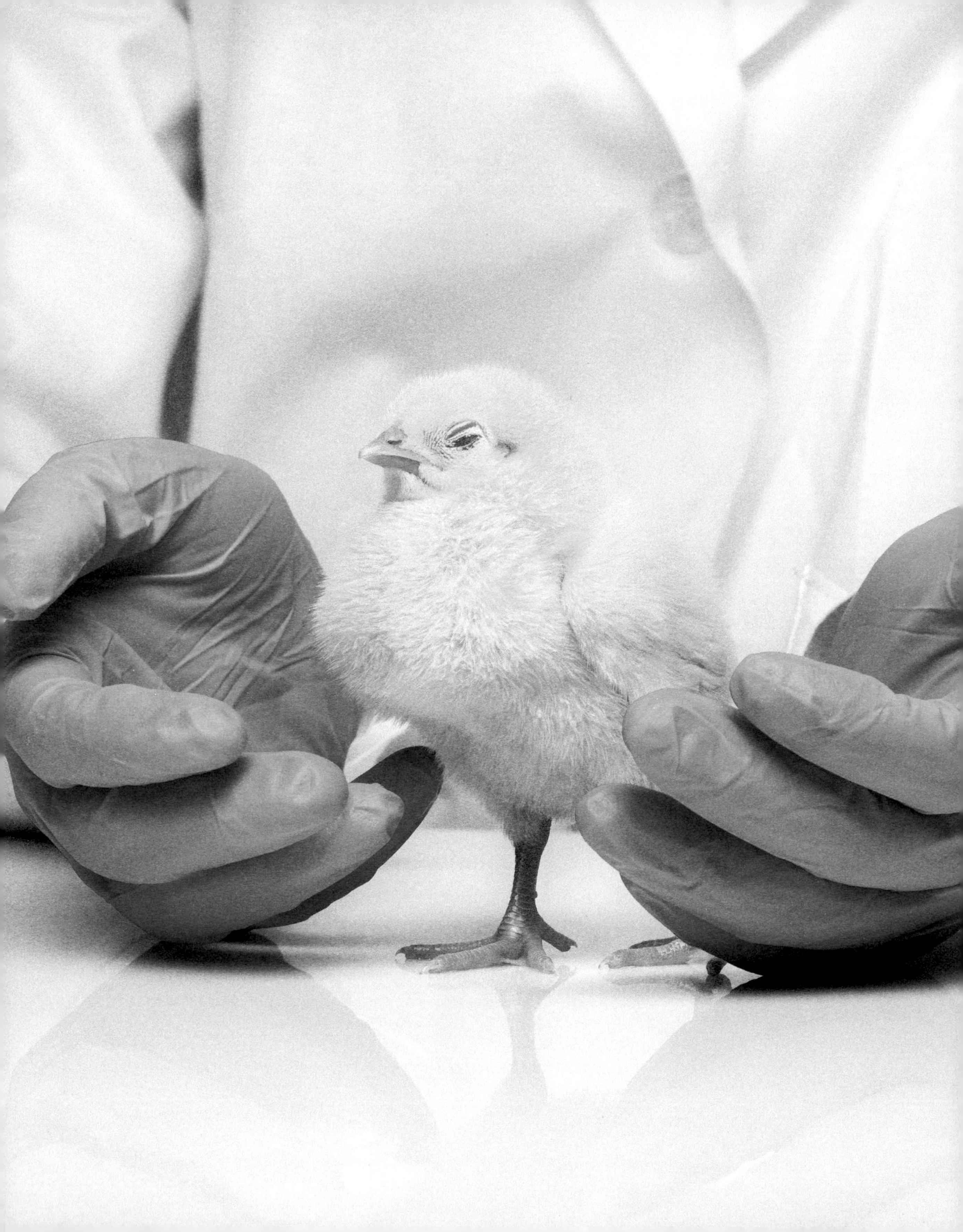

Illness and Ailments: Prevention and Treatment

"Pumpkin seeds are an all-natural dewormer in chickens."

☐ FACT ☐ POOP

TOPIC

ILLNESS AND AILMENTS NO. 1

ANSWER PROVIDED BY

Maurice Pitesky, DVM, MPVM, DACVPM
Veterinarian/Assistant Specialist in Cooperative Extension Poultry Health and Food Safety Epidemiology
School of Veterinary Medicine
University of California, Davis

The research

At some point, most chicken owners have to deal with internal poultry parasites or worms, which are very common. But not all worms are created equal, meaning no single dewormer that can kill all the different types of worms that can infect chickens. For example, while the drug piperazine may be effective against roundworms, it is not effective against other worms, including capillary worms. Therefore, when you see a statement indicating that pumpkin seeds are an all-natural dewormer, your first question should be, what type of worms? Chickens can get roundworms, capillary worms, cecal worms, tapeworms, and gapeworms.

In addition, it is important to consider the active ingredient and what the mode of action is or is presumed to be to understand at some level how the parasite is killed. For example, piperazine targets the nervous system of roundworms and paralyzes them, which causes them to detach from the intestinal wall of the GI tract. Once detached, the roundworms can be flushed out of the bird.

So now, let's move on to pumpkin seeds. Looking at the scientific literature, a 2016 article by Matthews et al. titled "Investigation of possible pumpkin seeds and ginger effects on gastrointestinal nematode infection indicators in meat goat kids and lambs" reviewed results of a study that naturally inoculated goat kids with gastrointestinal nematodes. Results, published in *Small Ruminant Research*, showed that pumpkin and ginger did not effectively reduce nematode levels.

"You would typically notice a *Salmonella* problem in your flock."

ANSWER PROVIDED BY

Megin Nichols, DVM, MPH, DACVPM,
Enteric Zoonoses Activity Lead Centers for Disease Control and Prevention

Chicken poop. Poultry usually appear healthy and clean when infected with *Salmonella*. *Salmonella* cannot be seen and may not cause illness in backyard chickens, ducks, and other poultry.

However, there has been other promising literature in this area. For example, pumpkin kernels and garlic may have some efficacy against one type of pathogenic nematode species in ruminants (Strickland et al., 2009, *Journal of Animal Production Science*). In addition, scientific studies have also explored the efficacy of using ginger to eradicate intestinal worms in cattle, horses, and lambs. However, to extrapolate from these studies from one species to another and from one species of internal parasite to another is not scientifically or practically appropriate or reliable.

The verdict

Mainly poop. Pumpkin seeds are not a natural dewormer.

What it means

A healthy chicken can have worms. If your chickens are healthy and you notice a worm or two, don't dwell on it. However, in general, eggs and poop should be the only things that come out of your chickens' vents.

"Backyard birds are more immune to the bird flu than factory birds."

☐ FACT ☐ POOP

TOPIC

ILLNESS AND AILMENTS NO. 2

ANSWER PROVIDED BY

Maurice Pitesky, DVM, MPVM, DACVPM
Veterinarian/Assistant Specialist in Cooperative Extension Poultry Health and Food Safety Epidemiology
School of Veterinary Medicine
University of California, Davis

The research

There are hundreds of different breeds of chickens distinguishable from one another by physical traits such as size, feather color, comb type, etc. Just as the birds look different, their subtle genetic differences have the potential to vary their productivity level and disease resistance.

One well-known example is the Fayoumi breed, which has some resistance to the protozoal parasite coccidia. However, Fayoumi are also less productive than commercial breeds and are somewhat notorious for their "high-strung" behavior. Consequently, disease resistance is not the only trait commercial and backyard producers should select for.

Alternatively, with the cost and ease of genetic sequencing, scientists can now identify genes that confer disease resistance, which can then be used in selective breeding programs with productive breeds that are easier to work with. However, the science on this is just beginning to emerge.

A study by Sironi et al. published in 2008 in the journal *Virology* showed two out of five chicken lines—White Leghorns and a commercial broiler line—had some resistance to experimental exposure to the H7N1 avian influenza virus. These types of results suggest that some chicken breeds have a natural resistance to influenza virus infection. However, even the "resistant" lines had mortality and the researchers were unable to identify the gene or genes responsible for the increased resistance. That last point is key to identifying both susceptible breeds and those with the potential for genetic crosses that produce birds with these types of genetic traits.

From an international food security perspective, these types of experiments are essential and will hopefully someday result in chicken breeds less likely to succumb to avian influenza and other diseases.

The verdict

Mainly poop. The blanket statement that backyard birds or even pastured birds are immune from avian influenza is inaccurate, in part, because there are no good scientific studies to demonstrate this claim. As Carl Sagan once said, "Extraordinary claims require extraordinary evidence."

What it means

Some people argued that during the 2014–15 highly pathogenic avian influenza outbreak in North America, high morbidity and mortality only affected conventional farms and, hence, backyard birds were resistant. While the data suggest that conventional flocks were most affected, this is most likely due to the fact that the majority of domestic poultry are conventional birds as opposed to free-range and backyard poultry. However, it's essential to maintain a wide genetic diversity of different breeds of backyard chicken in the hopes of eventually identifying unique traits—including disease resistance—in poultry.

"I should use antibiotics in my backyard poultry."

ANSWER PROVIDED BY

Megin Nichols, DVM, MPH, DACVPM,
Enteric Zoonoses Activity Lead Centers for Disease Control and Prevention

It depends. Proper vaccination, good biosecurity, good management, proper sanitation practices, and good nutrition are important to prevent disease in poultry and are far more effective than using antibiotics. Certain antibiotics are prohibited for extra-label use in food animals, including backyard poultry; even if you consider your backyard chickens pets, backyard poultry are still technically food-producing animals. In some instances, antibiotics may be useful to treat and control disease. A veterinarian should be consulted; judicious use of antibiotics may be recommended after appropriate testing.

Using antibiotics creates the potential for resistant bacteria to survive and multiply even when other bacteria are being killed. Antibiotic resistance has increasingly become one of the world's most pressing public health problems. Currently, many bacteria are able to resist the effects of an antibiotic, meaning the bacteria are not killed and their growth is not stopped. There is strong evidence provided by scientists around the world that shows that the use of antibiotics in food animals, such as chickens, can lead to resistant infections in humans. Even types of poultry not being used for food can still pass on antibiotic-resistant infections through the same ways non-antibiotic-resistant infections are passed from animals to humans.

"Marek's vaccine causes Marek's disease."

☐ FACT ☐ POOP

TOPIC

ILLNESS AND AILMENTS NO. 3

ANSWER PROVIDED BY

Maurice Pitesky, DVM, MPVM, DACVPM
Veterinarian/Assistant Specialist in Cooperative Extension Poultry Health and Food Safety Epidemiology
School of Veterinary Medicine
University of California, Davis

The research

Chickens vaccinated against Marek's disease rarely get Marek's disease. If they do, it's because a more virulent strain of the virus infected the chicken and caused disease. The vaccine exposes the chicken's immune system to a related nonpathogenic virus.

However, it should be pointed out that while Marek's vaccines effectively prevent Marek's disease and the resulting tumor formation, the vaccine is not considered "sterilizing" and doesn't prevent infection or shedding of the Marek's disease virus ubiquitous in the environment.

In addition, though the vaccine doesn't cause the disease, vaccinated birds appear more likely to transmit the virus, according to a 2015 study in the journal *PLOS One* by Read et al., "Imperfect Vaccination Can Enhance the Transmission of Highly Virulent Pathogens." Specifically, the investigators looked at two groups of birds: one vaccinated and one not. They exposed both vaccinated and unvaccinated birds to a virulent strain of the Marek's disease virus and then exposed each to two, new unvaccinated groups of chickens. Unvaccinated birds from the second group exposed to the vaccinated birds had the highest mortality. This virus-boosting phenomenon is known as "imperfect vaccine hypothesis."

The verdict

Mainly poop. The Marek's vaccine does not cause the disease in chickens that receive it. However, their vaccination may affect other, unvaccinated birds.

What it means

Big picture, Marek's disease is believed to be the most common mortality source in backyard poultry with no history of vaccination. Vaccinating your birds either by the hatchery in ovo (i.e., at approximately day 18 in the developing embryo) or at day 1 of age subcutaneously are the best ways to prevent your chickens from getting Marek's disease. You should also remove feathers from your coop—feather follicles are highly infective—before introducing chickens.

“Diatomaceous earth is an all-natural treatment for external parasites. It works by making contact with the waxy outer cuticle of insects, causing death.”

FACT POOP

TOPIC

ILLNESS AND AILMENTS NO. 4

ANSWER PROVIDED BY

Maurice Pitesky, DVM, MPVM, DACVPM
Veterinarian/Assistant Specialist in Cooperative Extension
Poultry Health and Food Safety Epidemiology
School of Veterinary Medicine
University of California, Davis

The research

There is a lot of research related to diatomaceous earth (DE) as a method to successfully control ectoparasites, or parasites that live on the surface of the body, like mites and lice. DE works by making contact and then absorbing the waxy outer cuticle of the insects, causing death by desiccation and, to a lesser extent, damage to the cuticle via the abrasiveness of the DE on the cuticle (Fields, 2000 and Quarles, 1992).

According to Amy Murillo, PhD, at University of Riverside in California, it is best to mix DE with sand (regular play sand works great) in a 1:4 ratio of DE to sand in a container such as a plastic swimming pool or cement mixing bin. Make sure to wear a dust mask while doing this since DE is safe for birds but can cause irritation in humans. This should attract hens to dustbathe in the mixture, which gets the DE into the feathers and onto the skin where the ectoparasites live. Having DE in the environment (i.e., on straw or dirt) does not appear effective on its own for ridding birds of ectoparasites.

There is far less information on using DE for control of endoparasites, or parasites that live inside the body of the host. In a scientific study by Bennett published in the journal *Poultry Science* in 2011, 2 percent of hens' diets were supplemented with food-grade DE (ie. DE with less than 7 percent composition of crystalline silica). Bovan Browns, a commercial brown feathered egg layer, fed dietary DE had significantly lower *Capillaria* and *Heterakis* worm burdens and fecal egg counts than the control Bovan Brown hens. However, no significant difference was noted for coccidia.

In the same study, Lowman Brown (LB) hens, a different commercial egg-laying brown chicken, showed no difference, even though LB hens are known to have genetic resistance to parasite infection. Consequently, it appears that the LB chickens didn't need the "help" that the DE may provide for internal parasite control.

The verdict

Mainly fact, but more research is necessary. In summary, there does appear to be some benefit and no effect on production to using DE. However, further studies are required to validate and expand on how best to use it.

"My birds don't need the Marek's vaccine. They are naturally immune."

☐ FACT ☐ POOP

TOPIC

ILLNESS AND AILMENTS NO. 5

ANSWER PROVIDED BY

Maurice Pitesky, DVM, MPVM, DACVPM

Veterinarian/Assistant Specialist in Cooperative Extension Poultry Health and Food Safety Epidemiology

School of Veterinary Medicine

University of California, Davis

The research

Unfortunately, there are no absolutes or perfect treatments in biology: for example, when someone with the flu sits between me and my wife on a plane, and I get the flu and she doesn't despite exposure to the same amount of virus. Her immune system might respond more robustly than mine, and I end up getting sick but she remains fine. For the same reason, blanket statements like "My birds don't need a Marek's vaccine because they are naturally immune" are just not true.

The Marek's virus is considered ubiquitous in poultry environments, meaning all poultry are considered exposed at day one of life, hence the value of early vaccination (e.g., in ovo at day 18 of embryonation by hatcheries with appropriate equipment, or day 1 of life by hatchery or owner). A combination of maternal antibodies passed from hen to chick and the above described vaccine schedule will significantly mitigate the prevalence of Marek's disease in your flock.

Alternatively, if you rely only on the maternal antibodies passed from hen to developing embryo, the chick is only protected for approximately the first three weeks of life. After that, the chick's immune system needs to produce antibodies. The value of vaccines in general is to help an animal produce a robust immune response by exposing it to a harmless antigen that "looks like" the harmful one. This scenario is preferable in that it primes the immune system for an attack by Marek's virus in the environment.

One aspect of the question implies that some birds don't need a Marek's vaccine because their genetics make them naturally immune. To that general point, there is some interesting work that looks at unique genetic sequences related to various parts of the chicken's immune system. The entire 1 billion DNA base pair chicken genome was sequenced in 2004. Those 1 billion base pairs are associated with between 20,000 and 25,000 genes (In comparison, the human genome has around 2.5 billion base pairs with approximately 20,000 genes). Identifying genes that play a role in the immune system and then further identifying unique sequences in some poultry associated with increased genetic resistance to Marek's disease and others is an active area of research. However, this approach should be seen not as a panacea but as another control measure to augment—not replace—current control measures like vaccination and proper husbandry practices such as removal of feather dander in coops with young.

The verdict

Poop. As I said, there are no absolutes or perfect treatments in biology. A combination of maternal antibodies and vaccinations is the best bet to safeguard chickens.

"You only get *Salmonella* from chickens that are sick with it themselves."

Chicken poop. Backyard poultry—including chickens, ducks, geese, and turkeys—can carry *Salmonella* in their intestines without appearing sick. It is common for live poultry to appear healthy and clean yet still be shedding germs through their droppings. These germs also commonly exist on their bodies (feathers, feet, and beaks) and on cages, coops, feed and water dishes, hay, plants, and soil in the areas where they live and are cared for.

ANSWER PROVIDED BY Megin Nichols, DVM, MPH, DACVPM,
Enteric Zoonoses Activity Lead Centers for Disease Control and Prevention

"Giving a Marek's vaccine booster will help the birds later in life against Marek's."

☐ FACT ☐ POOP

TOPIC

ILLNESS AND AILMENTS NO. 6

ANSWER PROVIDED BY

Maurice Pitesky, DVM, MPVM, DACVPM
Veterinarian/Assistant Specialist in Cooperative Extension Poultry Health and Food Safety Epidemiology
School of Veterinary Medicine
University of California, Davis

The research

The virus that causes Marek's disease is so ubiquitous in the environment (e.g., infection of chicks can occur almost immediately after hatching) that ideal vaccination occurs either in ovo on the 18th day of incubation or by subcutaneous administration of vaccine at one day of age.

Several types of Marek's vaccination can be given in ovo or subcutaneously. The in ovo vaccination method is more mechanized and, hence, more efficient—but only if the hatchery has the proper vaccinating machinery. This is typically done in hatcheries that provide chicks to commercial broiler farms. The additional advantage of the in ovo vaccine is that it is given approximately four days before the day one of life with the Marek's vaccine, meaning there is increased time between vaccination and exposure. This allows the immune system more time to develop an immune response before being challenged by Marek's virus in the environment.

Finally, it is important to realize there are different types of Marek's vaccines. Commercially, it is common practice to combine several vaccine types in an effort to broaden protection. At the minimum, use a lyophilized, or powdered version reconstituted with sterile diluent, vaccine at day 1 of age. The herpesvirus of turkeys (HVT) vaccine, a good starter vaccine, is commonly available from many feed stores. If one of your chickens dies and a diagnostic laboratory confirms that the cause is Marek's, consider using other types of Marek's vaccination on future flocks in your coop environment.

One other point: For the most part, Marek's disease is most common in birds younger than eighteen weeks of age. There are other tumor-forming viruses that cause Marek's-like symptoms that are more common in older birds. Unfortunately, the only diagnostic tests we have are at necropsy. However, getting a necropsy is important because identifying the cause of death is essential toward protecting the remainder of your flock.

The verdict

Mainly poop. Big picture, a booster shot will not hurt your chickens, but it will not help either, meaning you are wasting money and time by vaccinating your birds against Marek's disease later in their lives.

That being said, vaccination against Marek's disease constitutes an outstanding example of successful disease control in veterinary medicine. It is only efficacious if the vaccine is done in ovo or subcutaneously within one day. After one day, those birds should be considered "vaccinated improperly." Work with your feed store or hatchery to make sure they give Marek's disease vaccines at the proper time.

"Because my chickens are healthy and clean, they do not carry *Salmonella*."

Chicken poop. Though your chickens may indeed be healthy, they can still be infected with *Salmonella*. Backyard poultry can also become infected with *Salmonella* through their environment, by eating contaminated food, or from their mothers before they are even born or hatched. *Salmonella* is naturally occurring in the intestines of many different animals, so when chickens shed it through their droppings, it is easy to contaminate their body parts and their environment.

ANSWER PROVIDED BY Megin Nichols, DVM, MPH, DACVPM,
Enteric Zoonoses Activity Lead Centers for Disease Control and Prevention

"Garlic, cayenne pepper, and black walnut tincture deworm chickens."

☐ FACT ☒ POOP

TOPIC

ILLNESS AND AILMENTS NO. 7

ANSWER PROVIDED BY

Maurice Pitesky, DVM, MPVM, DACVPM
Veterinarian/Assistant Specialist in Cooperative Extension Poultry Health and Food Safety Epidemiology
School of Veterinary Medicine
University of California, Davis

The research

Internal poultry parasites or worms are very common, and at some point, most chicken owners have to deal with them. Also, all worms are not created equal, meaning that, unfortunately, no single dewormer can kill all the different types that can infect chickens.

To the claim above, there is no scientific literature supporting the efficacy of cayenne pepper and/or black walnut tincture in protecting poultry against internal parasites. Garlic, in contrast, has been well studied.

One of the active ingredients in garlic, allicin, has been found to be antibacterial, antiviral, antifungal, and antiparasitic. However, despite some evidence of garlic's efficacy against infectious agents, do no interpret this to mean that garlic is efficacious against all bacterial, viruses, fungi, and parasites.

For example, in a 2011 *Poultry Science* article by Velkers, the investigators studied the efficacy of commercially available garlic products that contained a high concentration of allicin against chickens experimentally infected with roundworms. Results showed no differences between infected chickens treated with the garlic product and infected chickens not treated (both groups had high roundworm loads upon necropsy). In contrast, infected chickens treated with the antiparasitic flubendazole had no roundworms in their intestines.

One side note: Flubendazole and its metabolites can subsequently end up in egg yolks, so it is important to work with a veterinarian or contact the Food Animal Residue Avoidance Databank (www.farad.org) to get the most up-to-date information on drug withdrawal times. I would also suggest contacting the FDA before giving a drug in general since waiting until after runs the risk of use of a drug for which there is no known withdrawal time. In that situation, the safest recommendation is to never consume any products from that chicken.

The verdict

Poop. In summary, there is no scientific literature supporting the efficacy of cayenne pepper and/or black walnut tincture in protecting poultry against internal parasites. And evidence shows that the active ingredient in garlic does not effectively deworm against common poultry parasites.

"Apple cider vinegar prevents coccidiosis and every other disease."

☐ FACT ☐ POOP

TOPIC

ILLNESS AND AILMENTS NO. 8

ANSWER PROVIDED BY

Maurice Pitesky, DVM, MPVM, DACVPM
Veterinarian/Assistant Specialist in Cooperative Extension Poultry Health and Food Safety Epidemiology
School of Veterinary Medicine
University of California, Davis

The research

Before we talk about apple cider vinegar as a potential treatment for the parasite coccidia and the disease coccidiosis, it is important to recognize that like all poultry diseases, the most bang for your buck comes from focusing on prevention/biosecurity with respect to coccidia control. In this case, that means focusing on controlling moisture where your poultry live, eat, and drink and feeding your chicks a starter feed medicated with a coccidiastat. (There is no equivalent human analog, so worries of antibiotic resistance, transmission, and human disease aren't a concern in this situation. This fact is reflected by the FDA's updated Veterinary Feed Directive [VFD], which allows the use of anticoccidial medications in poultry without the permission of a licensed veterinarian.)

Organic starter feed does not contain coccidiastat. Therefore, pay even more attention to the birds' environment to avoid high loads of coccidia in the soil and substrate on which the birds are raised. That being said, you can never rid an environment entirely of coccidia; and interestingly, low loads of it in the environment actually expose birds to low doses that can act as a "vaccinating dose."

Now, let's address apple cider vinegar. For some reason, it's often considered a cure-all/panacea for a variety of maladies in humans and animals alike—for everything from hiccups, diabetes, and constipation in humans to improved digestion and arthritis in horses to increases in egg production and infectious disease treatment in poultry. This would suggest that vinegar added to drinking water and feed and applied topically is the ticket to solving nearly any problem.

Some of it can be explained by the history of food preservation, in which vinegar was made from multiple perishable products, including apples via fermentation: That process—plus vinegar's 5,000-year history.

But scientists and veterinarians primarily focus treatment decisions on evidence-based medicine in peer-reviewed scientific literature. To date, no studies show any benefit from apple cider vinegar for treating coccidia in any species, including poultry.

The verdict

Poop. Does the fact that there is no scientific literature on the subject mean it absolutely doesn't work? No, but it is important for consumers to be skeptical of claims that currently appear to have no foundation in science. Ask for evidence and email, call, and visit experts—including your friendly university cooperative extension specialists!

"People get sick from *Salmonella* because they have no immunity to it."

Chicken poop. People get sick from *Salmonella* because they are directly or indirectly exposed to an animal, food source, or surface that is harboring the bacteria. Immunity to *Salmonella* bacteria does not develop over time or after a certain number of infections, but having a stronger immune system can make a person less susceptible to getting sick. The elderly, infants, and people with weakened immune systems are at a higher risk of infection and severe illness, especially children under five years of age.

ANSWER PROVIDED BY Megin Nichols, DVM, MPH, DACVPM,
Enteric Zoonoses Activity Lead Centers for Disease Control and Prevention

“You can always get a complete diagnosis of your sick chicken by sending a photo to your veterinarian.”

FACT POOP

TOPIC

ILLNESS AND AILMENTS NO. 9

ANSWER PROVIDED BY

Maurice Pitesky, DVM, MPVM, DACVPM
Veterinarian/Assistant Specialist in Cooperative Extension Poultry Health and Food Safety Epidemiology
School of Veterinary Medicine
University of California, Davis

The research

I once heard someone at a meeting say to an audience member, "Send me a picture of your sick chicken, and I can help diagnose it," that this was a better alternative than working with a diagnostic lab. Pictures can be somewhat useful in moving toward an initial diagnosis but can provide definitive answers for just a few diseases.

For example, pictures of bloody diarrhea or a chicken with nasal discharge don't lead to what veterinarians call "pathopneumonic" or definitive identification of a specific disease. In addition, many diseases require a necropsy, the veterinary term for autopsy, during which a trained veterinary pathologist looks at the internal organs to better identify what disease (i.e., infectious vs. noninfectious) is occurring. Further microscopic analysis by histopathology, various other types of microscopy, and isolation of virus, bacteria, fungi, and parasites are often necessary to make a final diagnosis.

The verdict

Mainly poop, though somewhat helpful: If it's too good to be true, it's probably not true. In other words, the diagnosis a picture reveals probably isn't the true or complete answer.

However, photographs can be useful for the identification of some diseases such as avian pox. Specifically, pink scabs or wart-like nodules, typically around 1 mm in size, or similar lesions across the comb, wattles, eyelids, and other nonfeathered areas are typical of the dry form of avian pox. As an extension veterinarian, I have worked with owners who sent me photographs of lesions similar to the ones described above, along with a complete history, and we've used that information to tentatively diagnose dry avian pox.

What it means

Veterinary pathologists go to veterinary school and do residencies in pathology so they can work in a diagnostic lab. It takes years for them to become experts in one aspect of pathology. Hence, large diagnostic labs have multiple pathologists with different levels of expertise to work collaboratively to identify disease cause. Most states have veterinary diagnostic labs associated with a university system that, in many cases, provide free or heavily discounted care with the hope that early disease diagnosis will prevent large-scale disease outbreaks. The services are typically easily available and cost effective, so use them if you need to.

Bottom line: Though pictures can be helpful, veterinarians are not allowed to prescribe medications based on a phone call and/or photograph. Veterinarians can only treat patients seen in person.

"Diagnostic labs will kill all your birds if you bring in animals found to be sick."

☐ FACT ☐ POOP

TOPIC

ILLNESS AND AILMENTS NO. 10

ANSWER PROVIDED BY

Maurice Pitesky, DVM, MPVM, DACVPM
Veterinarian/Assistant Specialist in Cooperative Extension Poultry Health and Food Safety Epidemiology
School of Veterinary Medicine
University of California, Davis

The research

Unfortunately, I heard this comment from a speaker at a poultry meeting. Before we talk about two scenarios where this can happen—if your birds are diagnosed with either the H5 or H7 version of the avian influenza virus or with the virulent Exotic Newcastle Disease (end)—we should discuss the overall mission of veterinary diagnostic labs.

Many states have diagnostic labs staffed by veterinarians with expertise in diagnostics of disease funded via public funds or who focus on diagnosing diseases of commercial and backyard livestock and poultry. The goal is simply to facilitate disease identification, to help producers and enthusiasts identify problems in their herd or flock. Once the diagnostic lab provides this information, it's up to the owner to then take actions.

Getting back to the statement, the idea that veterinarians will "kill all your birds" is so ludicrous that had I not heard the statement in person, I would doubt its authenticity or at least believe there was some type of caveat to the statement. There was not. In fact, the speaker doubled down on the statement, saying he could help diagnosis disease by simply talking to the owner and looking at a photograph or two. The idea that this approach will result in a more accurate diagnosis than pathology, virology, bacteriology, histopathology, and electron microscopy is simply not true.

However, the two diseases mentioned in the first paragraph will result in the depopulation of your flock. This is for good reason because these diseases are considered foreign-animal diseases that currently have no treatment. They're also dangerous and easily spread between flocks. It should be noted that these diseases are very rare "newsworthy events" when they happen. More typical are *E. coli* or coccidiosis. That being said, if your birds do have any disease, especially avian influenza or END, you should want to know that and respond in such a way that prevents spread to other flocks. This is part of being a responsible poultry owner.

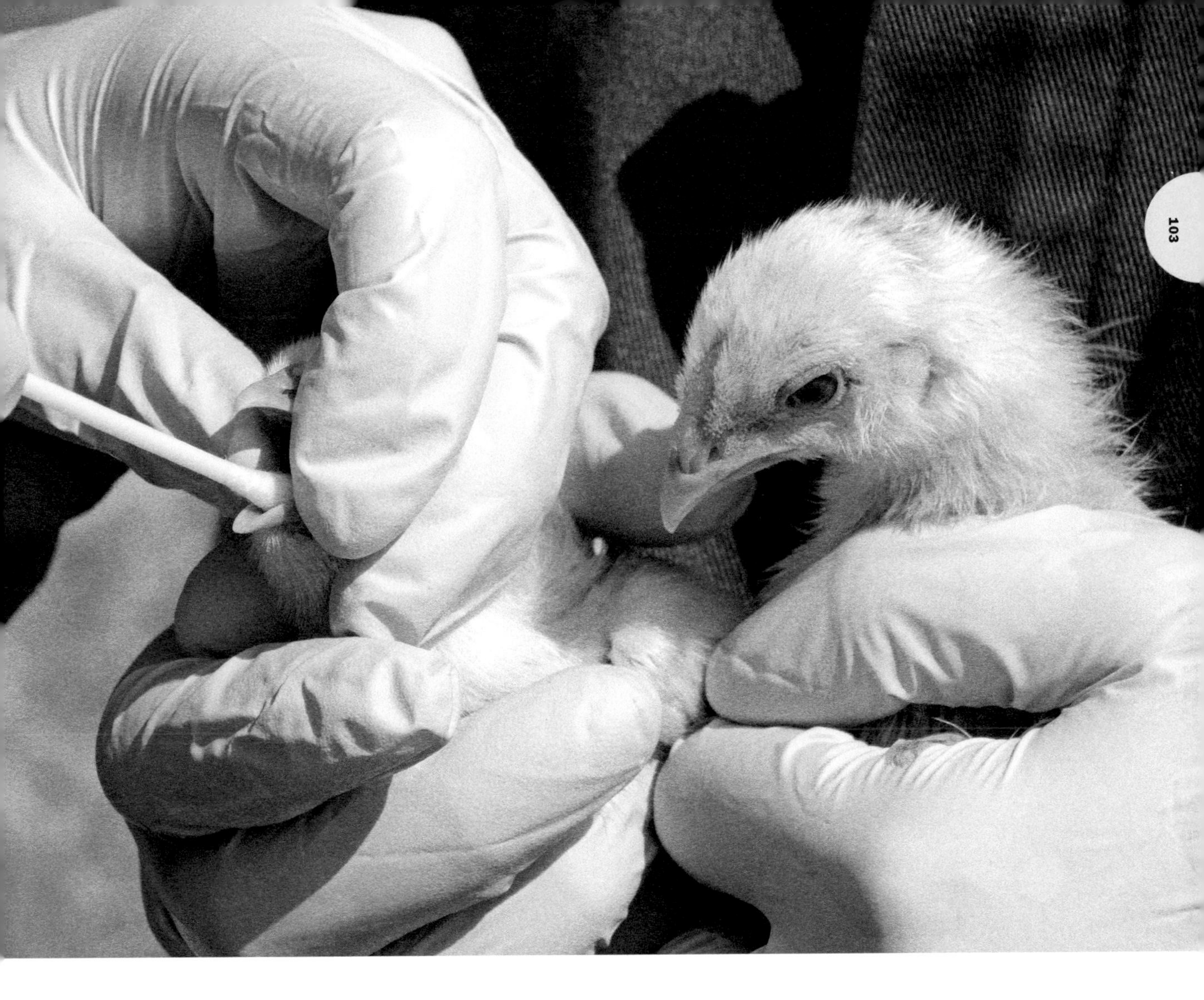

The verdict

Poop. Diagnostic labs aim to diagnose disease using techniques like pathology, virology, bacteriology, histopathology and electron microscopy. They do not aim to kill off sick birds, unless the chickens test positive for the H5 or H7 version of the avian influenza virus or for END.

What it means

It's fine to bring your birds to state diagnostic labs. However, there are other option; for example, working with private veterinarians or public veterinarians based at land grant universities throughout the country (and whose consultation is free) to seek the best overall management option.

“Flea medications such as FRONTLINE and Advantage are OK to use in poultry.”

FACT POOP

TOPIC

ILLNESS AND AILMENTS NO. 11

ANSWER PROVIDED BY

Maurice Pitesky, DVM, MPVM, DACVPM

Veterinarian/Assistant Specialist in Cooperative Extension Poultry Health and Food Safety Epidemiology

School of Veterinary Medicine

University of California, Davis

The research

"I've used over-the-counter flea products on my chickens to kill fleas or mites, and they work great."

I've heard that statement from backyard poultry enthusiasts at meetings, read it on poultry blogs, and even seen it on YouTube. The statement is correct in that the active ingredients in FRONTLINE and Advantage—fipronil and imidacloprid, respectively—can be effective against a wide range of ectoparasites including fleas, ticks, and mites, depending on the drug used. However, chickens are food animals, so you need to consider more than just drug efficacy. For example, how is the drug metabolized and how long can it persist in eggs and muscle? Because of these concerns, flea medications are not recommended for use in poultry.

In addition, the Food Animal Residue Avoidance Databank (FARAD), which provides science-based advice related to withdrawal intervals for food animals, advises that FRONTLINE cannot legally be used in an extra-label manner because it is not an FDA-approved product, a requirement for such a use. In regards to Advantage, not enough pharmacokinetic data exist to make a blanket withdrawal interval recommendation for poultry products.

Ectoparasites are "ecto," meaning they live on the skin and can be identified without a necropsy. This is important because, like all diseases, identification of the cause—in this case the type of ectoparasite—is key to achieving pest control.

The verdict

The answer to this is 100 percent poop! Do not use any type of flea medication on your chickens.

What it means

If you're trying to figure this out on your own, stick to reputable resources like universities. Alternatively, if you don't want to become an amateur entomologist, submit samples to a diagnostic lab. Once the disease is identified, you can work with your local feed store, veterinarian, or university-based extension veterinarian to determine the best treatment and preventative measures.

Also, remember to continuously monitor your birds' health. This includes a thorough check of your chickens' feathers from head to the vent. Just like in humans, the earlier a problem gets identified, the easier it is to resolve.

"Chicks raised outdoors won't have diseases as they age because they were kept naturally."

☐ FACT ☐ POOP

TOPIC

ILLNESS AND AILMENTS NO. 12

ANSWER PROVIDED BY

Maurice Pitesky, DVM, MPVM, DACVPM

Veterinarian/Assistant Specialist in Cooperative Extension Poultry Health and Food Safety Epidemiology

School of Veterinary Medicine

University of California, Davis

The research

This statement reminds me of the "hygiene hypothesis," which suggests that one explanation for the increase in allergic diseases in modern times is the relative lack of exposure to infectious agents in young children.

Specifically, the hypothesis suggests that in order for the immune system to develop appropriately, it needs to come into contact with a variety of microorganisms during infancy. Because many infants in the developed world grow up in environments with fewer microorganisms, including those associated with farm animals and plants, the theory goes that their immune systems have not been adequately challenged. So as they grow older, their immune systems respond inappropriately to environmental allergens. This leads to increased allergy rates.

I won't spend the remainder of the space here arguing for or against the hygiene hypothesis. However, in chickens, scientific support exists for exposing birds to microorganisms, including low loads of some disease-causing agents. In essence, this describes a "vaccinating dose."

For example, low loads of coccidia in the soil are believed beneficial for flock health and protection against coccidiosis. However, larger loads of coccidia are still considered detrimental. I once read an article in which an outdoor producer mentioned that his birds never need vaccines or antibiotics because of the "regenerative soils" on his farm. He may be describing this phenomenon. However, the idea that healthy soil offers absolute protection against all disease-causing agents in poultry is unfortunately not true.

The verdict

Poop. Given the hygiene hypothesis and the accurate notion that exposure to low levels of some pathogens and other microorganisms can be beneficial, it is easy to see why people might believe that birds raised outdoors won't have any disease when they get older. Unfortunately, that is not true. As an extension veterinarian, I can vouch for the fact that older birds raised outdoors do get disease. However, as all birds age, they are less likely to have disease; many, like coccidia, infectious bursal disease, and most types of Marek's disease, affect younger birds, both those raised outdoors and those raised indoors.

"Chickens don't get the same diseases as wild birds, so I don't need to worry about free-ranging them."

☐ FACT ☐ POOP

TOPIC

ILLNESS AND AILMENTS NO. 13

ANSWER PROVIDED BY

Maurice Pitesky, DVM, MPVM, DACVPM
Veterinarian/Assistant Specialist in Cooperative Extension Poultry Health and Food Safety Epidemiology
School of Veterinary Medicine
University of California, Davis

The research

The answer to this is "yes" and "no," but I will call it mainly "poop" because the idea that birds don't get any of the same diseases that wild birds do is a dangerous way to think with respect to your flock's overall health.

In general, there is a whole list of viruses and bacteria with known wildlife animal carriers, such as rodents and birds. Bacteria such as *Pasteurella multocida*, also called fowl cholera, and *Haemophilus paragallinarum*, also called infectious coryza, are two that can be present in both avian wildlife and poultry. Avian influenza and infectious bursal disease virus are known viruses present in avian wildlife that can then affect poultry. This is the primary reason veterinarians and other professionals who work with poultry focus on biosecurity.

Specifically, preventing contact and exposure of disease between wildlife and domestic poultry is the most important way to protect your flock from infectious diseases. Without going into too much detail, focus on putting up appropriate fencing and netting and eliminating magnets such as leaving feed in a place where wildlife will attempt to access it.

Unfortunately, treating poultry infected with these types of organisms is often less than satisfactory because often poultry are asymptomatic carriers and have the potential to expose other birds in your flock without being sick themselves. This is especially true in mixed-age flocks in which you periodically introduce younger birds that subsequently get infected with agents that older birds carry.

The verdict

Mainly poop. It's possible for your chickens to get the same diseases as wild birds. The best measure against this is appropriate biosecurity for your flock.

"Keep vitamins and electrolytes in the water at all times so your chickens will have stronger immune systems."

☐ FACT ☐ POOP

TOPIC

ILLNESS AND AILMENTS NO. 14

ANSWER PROVIDED BY

Maurice Pitesky, DVM, MPVM, DACVPM
Veterinarian/Assistant Specialist in Cooperative Extension Poultry Health and Food Safety Epidemiology
School of Veterinary Medicine
University of California, Davis

The research

We often take water for granted, but having access to clean, cool water 100 percent of the time is essential for your flock's overall health. This is especially true during the summer, when water intake can increase by up to three times the amount as other times of year.

In general, vitamins and electrolytes play a significant role in the metabolic functions of poultry. The National Research Council, which has developed suggested nutritional guidelines for various species of animal feed, suggests poultry get fourteen amino acids, twelve minerals, thirteen vitamins, and one fatty acid. Commercial feed companies typically closely follow these guidelines, as well as peer-reviewed literature on the subject.

My general inclination regarding nutrition in backyard poultry is to keep it simple. Use a commercial feed that is appropriate for your flock (e.g., layer versus broiler, chick versus layer). Only consider additions or changes when a problem arises. The feed already includes vitamins like A, D, and E; and drinking water naturally contains electrolytes like sodium, potassium, and chloride. Therefore, my general recommendation is, "Don't fix it if it ain't broke."

That being said, many commercial producers do add vitamin and electrolyte premixes to their poultry diet in part because the peer-reviewed literature suggests some benefits, particularly with respect to avoiding heat stress and overall production. However, it is important to do this under the supervision of a veterinarian or nutritionist because more does not always mean better. Fat-soluble vitamins like A, D, and E can accumulate and cause disease associated with toxicity. For example, Vitamin D3 toxicity can result in delayed crop emptying and regurgitation, lethargy, blood-tinged feces, dehydration, and kidney damage.

One time you *will* want to make a change is when you add either antibiotics or vaccines to your flock's drinking water, especially if the water source is

municipal. Municipal water contains a small amount of chlorine to control for pathogens like *E. coli*, but that same chlorine can inactivate vaccines and antibiotics. In this case, it's important to use water without chlorine, such as distilled water, or use commercial stabilizers to deactivate the chlorine. In addition, the antibiotic or vaccine is often diluted initially in reconstituted powdered milk to help buffer and protect the antibiotic and vaccine once it is added to the de-chlorinated water.

One other point: If you use well water, have it tested for bacteria and heavy metals like lead and mercury. Because chickens are food animals, their consuming these heavy metals means they could end up in eggs, organs, and meat.

The verdict

Poop. Unless there is a problem, stick with normal cool and clean water provided twenty-four hours a day, seven days a week. Only consider adjusting when necessary.

"Getting *Salmonella* is no big deal."

ANSWER PROVIDED BY

Megin Nichols, DVM, MPH, DACVPM,
Enteric Zoonoses Activity Lead Centers for Disease Control and Prevention

Chicken poop. *Salmonella* is an illness that causes one million illnesses in the United States each year, including 19,000 hospitalizations and 380 deaths. *Salmonella* infections can cause diarrhea, fever, and abdominal cramps that typically last four to seven days. Although most people recover without treatment, the resulting diarrhea can be severe enough that hospitalization is required. In some people, *Salmonella* infection can become life threatening when the bacteria spread from the intestines to the bloodstream and beyond. The elderly, infants, and immunocompromised individuals are more likely to have a severe illness.

“Giving probiotics to baby chicks reduces the risk of *Salmonella* in eggs and other diseases later in life.”

☐ FACT ☐ POOP

TOPIC
HOME SWEET HOME NO. 15

ANSWER PROVIDED BY
Brigid A. McCrea, PhD
Extension Specialist
4-H Youth Development Animal Programs
Alabama Cooperative Extension System
Auburn University

The research

There have been claims stating that giving chicks probiotics will prevent other diseases later in life. These claims are occasionally science based but only partly true. Most of the research being referenced is done on broiler chickens, short-lived meat-type chickens raised to just six to eight weeks of age, in conventional housing. Given their lifespan and purpose, slight improvements to their immune system, particularly in the gut, can yield a viable dollars-and-cents result for companies.

The good news is that it appears there is potential for real and true change to the gut of the chicken. Researchers at universities with strong poultry programs have been working on probiotic and direct-fed microbials to determine their effects on production performance in broilers (Haghighi et al., 2006, *Clinical and Vaccine Immunology*). However, because research to date has only occurred in broilers, the results do not yet extend to the living systems and breeds associated with backyard poultry. At this point, there are just too many factors to control for, and it would take a systematic approach to determine which, if any, affect the probiotics in either a positive or negative way.

As to the question of whether these probiotics and direct-fed antimicrobials can help against *Salmonella*, it's been shown that certain bacteria found in them stimulate the immune system and improve the growth of young chickens bred for meat production (Higgins et al., 2007, *Poultry Science*). That being said, even a probiotic that works to reduce *Salmonella* in the gut will not necessarily reduce *Salmonella* in the egg and vice versa—though the correlation between a chicken with *Salmonella* in the gut and laying *Salmonella*-infected eggs can be high occasionally.

Many moving parts help determine whether salmonellae take up residence in a hen's liver and cause her to lay infected yolks. The gut is ever changing, so every day brings a new situation for the digestive system of a bird with access to the outdoors. Depending on the preparation of the immune system, the gut may or may not be able to tackle what the bird throws at it each day. Even commercial poultry production houses test for *Salmonella* with great regularity because they realize that factors change, which can cause even birds that weren't sick yesterday to become sick.

The verdict

This statement is too broad. The research is still in progress, though some preliminary results indicate that along with excellent biosecurity measures, probiotics might be an ancillary measure to control *Salmonella* in the gut of conventional, commercially raised laying hens.

This does not, however, translate to laying hens or dual-purpose breeds. Many small flocks are raised outdoors, a scenario research does not often test. To make the jump between broilers and laying hens is far too great right now.

Plus, probiotics, and their effectiveness, have not been tested on every disease-causing organism. They are not likely effective against all parasites or viruses, both of which can affect birds later in life. The gut is always in flux as the bird gets exposed to new factors every day. Giving probiotics to a chick at a young age does not guarantee it immunity when older.

What it means

To take the above myth as fact, not only would we need more research about laying hens, but we would need to get more specific about whether the research refers to the gut or the eggs themselves.

Hopefully, future research will determine which probiotics are most effective against *Salmonella* and other organisms when the birds are young and when they are older, maybe even after their egg-producing prime. But we aren't there yet.

"If I get *Salmonella* once, I am immune to getting it again."

ANSWER PROVIDED BY

Megin Nichols, DVM, MPH, DACVPM,
Enteric Zoonoses Activity Lead Centers for Disease Control and Prevention

Chicken poop. Getting infected once does not create immunity to *Salmonella*. A person can become infected with *Salmonella* any number of times. Prevention, including washing hands with soap and water and cooking food to the proper temperatures, is the best way to avoid getting sick.

"Using kitchen/ garden waste will reduce my feed bill, but I should only feed them veggies right?"

ANSWER PROVIDED BY

Answer provided by Zac Williams, Ph.D.

Independent Poultry Scientist

Chicken poop. Chickens either wild or allowed to free range will eat almost anything they can find, including small rodents, animal carcasses, and so on. For some of you who may have had the misfortune of having a crowded coop, you know that they will eat other chickens also. Chickens, like all animals, need protein—they break it down to obtain the amino acids and use them for many essential body functions.

Of particular interest to poultry is the amino acid methionine. Methionine is one of the limiting amino acids in poultry, meaning they will require it in the diet more than any other amino acid. Poultry—chickens and turkeys especially—can actually crave methionine and will sometimes engage in feather picking if they are methionine deficient.

"Tomato juice acts as an electrolyte for chicks."

☐ FACT ☐ POOP

TOPIC

HOME SWEET HOME NO. 16

ANSWER PROVIDED BY

Brigid A. McCrea, PhD
Extension Specialist
4-H Youth Development Animal Programs
Alabama Cooperative Extension System
Auburn University

The research

There is no specific research stating that tomato juice is a suitable electrolyte to feed to chickens. For tomato pomace, however, some research does exist.

Tomato pomace is the leftover material from tomato processing, and it has been fed to laying hens. Researchers have tested the growth, mortality, and eggs in post-molt hens given tomato pomace. This particular research study was trying to use tomato pomace to see if its inclusion in the diet did or did not help the birds in any particular way as they were molting. Body weights, mortality, and eggshell thickness between hens fed a diet containing tomato pomace and hens fed a control diet ended up the same. The difference showed up in egg production: Fewer eggs came from hens fed a diet that included tomato pomace.

Similar research (Patwardhan et al., 2011, *The Journal of Applied Poultry Research*) was done to determine how effective tomato pomace might be in the bone health of molting birds. As it turns out, it does work well—but with the caveat that calcium should be added to the diet to prevent bone health problems. Keep in mind, the research with tomato pomace has been with commercial strains of single comb White Leghorn hens and took place in commercial cages.

The verdict

Poop. There is no information at this time to verify that tomato juice offers your chicks a suitable source of electrolytes. Most feed stores across the country do carry electrolyte and vitamin packs designed with poultry in mind. (These packages contain instructions on how to mix the electrolytes properly so you do not harm your flock.)

"The egg withdrawal period for Wazine is two weeks."

☐ FACT ☐ POOP

TOPIC

ILLNESS AND AILMENTS NO. 17

ANSWER PROVIDED BY

Ronette Gehring, BVSc, MMedVet (Pharm), Dipl. ACVCP

Department of Anatomy and Physiology
Institute of Computational Comparative Medicine
College of Veterinary Medicine
Kansas State University

The research

Wazine, an over-the-counter deworming medication, has a two-week withdrawal time, according to the FDA-approved label. However, that is specifically for meat, not eggs, and only for chickens and turkeys for which the drug has been administered according to label instructions.

This FDA-approved withdrawal time for meat cannot be extrapolated to eggs produced by backyard chickens for two reasons: differing disposition of the active ingredients and dosage.

Disposition of active ingredients has to do with the rate at which they distribute to and deplete from meat and eggs. This may differ for each food product, which can result in higher drug concentrations in eggs, resulting in the need for a longer withdrawal time.

In terms of dosage, the higher the dose, the longer the withdrawal time needed. Since Wazine is administered through a chicken's food or water, the actual dose that a backyard chicken receives may differ from the dose on which the FDA-approved label withdrawal time is based. When establishing the FDA-approved label withdrawal time, the source and amount of water and food can be tightly controlled to ensure a specific dose. But backyard chickens can obtain food and water from different sources, and their intake varies depending on several physiological, environmental, and behavioral factors.

The verdict

Mainly poop. The use of a product such as Wazine in backyard egg-producing chickens constitutes extra-label use and can only occur with the guidance of a veterinarian, who is responsible for ensuring that the withdrawal time is adequately extended to account for differences in drug dose and disposition.

“If you don’t eat the eggs or meat, it is okay to use antimicrobials such as enrofloxacin in poultry.”

☐ FACT ☐ POOP

TOPIC

ILLNESS AND AILMENTS NO. 18

ANSWER PROVIDED BY

Ronette Gehring, BVSc, MMedVet (Pharm), Dipl. ACVCP

Department of Anatomy and Physiology
Institute of Computational Comparative Medicine
College of Veterinary Medicine
Kansas State University

The research

Enrofloxacin is a fluoroquinolone, which is a group of antimicrobials that play an important role in treating bacterial infections in humans. There is ongoing concern about severe and life-threatening diseases becoming untreatable because the causal bacteria are developing resistance to the antimicrobials used to treat them. Therefore, steps have been taken to limit the use of these in food-producing animals.

This is not due to concerns about drug residues in the meat or the eggs but rather because of concerns about exposing bacteria in the animals’ guts to these antimicrobials. This exposure of the bacteria to the medication can result in resistance by killing susceptible bacteria and leaving the resistant ones to dominate the population. Passing on these resistant bacteria to humans, either through contaminated meat or eggs or through direct contact with the animals and their feces, could lead to untreatable infections in humans.

To minimize bacterial exposure to medically important antimicrobials, the Food and Drug Administration Center for Veterinary Medicine prohibits the extra-label use of fluoroquinolones in food-producing animals. All products containing fluoroquinolones that were historically approved by the FDA for treating chickens have been withdrawn from the market, making the use of enrofloxacin in this species illegal.

The verdict

Complete poop. It is never acceptable, and in fact, is now illegal, to use fluoroquinolones like enrofloxacin in any type of food-producing animals, including chickens, even if the animals’ meat and eggs do not get consumed. This includes all chickens, whether they are in a commercial hen house, part of a small backyard flock, or an individual pet.

"Garlic water and vitamin E oil cure blackhead."

☐ FACT ☐ POOP

TOPIC

ILLNESS AND AILMENTS NO. 19

ANSWER PROVIDED BY

Ronette Gehring, BVSc, MMedVet (Pharm), Dipl. ACVCP
Department of Anatomy and Physiology
Institute of Computational Comparative Medicine
College of Veterinary Medicine
Kansas State University

The research

This is another myth that appears true on the surface but begins to unravel as you learn more. It is always the hope that a natural replacement to chemical antibiotics be able to solve the problems associated with certain diseases. Many backyard chicken flock owners try to avoid the use of antibiotics, sometimes to the detriment of the sick bird. However, this myth may have some hope for flock owners who have a chicken correctly diagnosed with *Histomonas meleagridis*, otherwise known as blackhead disease.

First off, *Histomonas meleagridis* is a single-cell parasite called a protozoa. It affects turkeys and some lines of chickens. The organism has become increasingly problematic in commercial chicken breeding flocks. The symptoms are painful for us to look at in a bird that has been afflicted. Swelling of the sinuses and liver tissue damage combine in many cases with the failure of the circulatory system, which gives rise to the name of the disease, blackhead.

There are commercially available products that have been developed to combat the organism that causes blackhead diseases in Europe. This was done in response to the changes in management protocols for poultry raised within the borders of the European Union. Some of these products have had research done in order to determine their efficacy.

Therein lies the stretch that leads to the myth. The hope of backyard flock owners is that by feeding garlic in the water and perhaps vitamin E oil, you could cure a chicken or turkey affected by blackhead disease. These commercial products use garlic, don't they? So why not put garlic in the water as the simpler solution? Well, the commercial products contain extracts of garlic in the chemical form of its most potent and active component, allicin. The allicin in the products is many times stronger than that of a bulb of garlic. The compound allicin, in its purest form, is treated as a chemical; and care should be taken in its storage and handling.

Vitamin E, also referred to as alpha-tocopherol, is an antioxidant. Many antioxidants do have a measure of effectiveness in the prevention of illness by supplying the immune system with the building blocks it needs to perform different functions. However, the dose of the disease-causing agent, in any type of disease, may be sufficient to overwhelm the immune system.

So, the timing of the treatment, the dose of the infective agent, the host's immune system readiness, and the environmental conditions in which the bird lives are all factors that require testing and measurement before a broad statement about the effectiveness of garlic and vitamin E oil on blackhead disease.

The verdict

Poop. The research has yet to be done to prove that garlic water and vitamin E oil are an effective measure against the disease. You are better off getting a diagnosis from a veterinarian and a prescription of antibiotics. The period of time that your ailing bird will suffer is likely to be lessened with these expedient measures set into motion.

"If I feed herbs regularly to my chickens, then it will prevent diseases in my flock."

FACT POOP

TOPIC

ILLNESS AND AILMENTS NO. 20

ANSWER PROVIDED BY

Ronette Gehring, BVSc, MMedVet (Pharm), Dipl. ACVCP
Department of Anatomy and Physiology
Institute of Computational Comparative Medicine
College of Veterinary Medicine
Kansas State University

The research

Thankfully, due to changes in public perception and regulations, some counties have taken antibiotics off the market, thereby opening the door for research with herbs. Usually, the incorporation of herbs directly in the diet is not potent or efficient enough to make any difference in the bird. Therefore, the more concentrated components of the active ingredients of herbs are researched instead.

Keep in mind that not all herbs, or their concentrated extracts, are going to aid in prevention in all ages or breeds. Some of these herbs may be best utilized for direct treatment of an organism that is causing illness in a bird. Parameters for efficacy have not all been worked out, so let's explore a few examples of what is out there now.

One of the most troublesome viruses worldwide is avian influenza, and it looks like a commercial preparation of echinacea may have efficacy against the virus (Pleschka et al., 2009). Barrenwort and a propolis extract are both showing that they can stimulate the immune system into action when challenged with The Newcastle disease virus (Kong et al., 2006). *Salmonella* levels in a test tube experiment with mint extract showed an approximately threefold reduction (Tassou et al., 2000).

In some research, a panel approach is taken. This takes several batches of chickens in small groups and tests different types of herbs on them to see what, if any, improvements there are when a bird is challenged with disease-causing organisms. This screening process has been done with coccidiosis, and several potential plant materials have shown promise (Youn and Noh, 2001). One of the next steps in this type of research is to determine what the minimum dosage is that will create the beneficial effect. Also, researchers must consider the delivery mode of such items. For example, some materials may be inactivated by the pelleting process if it is incorporated into the feed.

The verdict

Mainly poop. As much as I would like to say that herbs will answer all the disease problems in the poultry world, such is just not the case . . . yet. So little research has been done so far that the door is wide open for efficacy trials. Just think about it: each combination of herb and disease organism and then throw in dosages, breeds, and age of bird. We have years of work ahead of us in the poultry research community.

"Getting *Salmonella* is just part of owning backyard poultry."

ANSWER PROVIDED BY

Megin Nichols, DVM, MPH, DACVPM,
Enteric Zoonoses Activity Lead Centers for Disease Control and Prevention

Chicken poop. Owning backyard poultry can be a positive experience and does not necessarily mean you will get *Salmonella*. However, it is important to consider the risk of illness and understand the precautions that can be taken to prevent infection. Live poultry, even when appearing healthy and clean, can shed *Salmonella* through their droppings, feathers, feet, and beaks. These germs can also be transmitted to surfaces or other items they come in contact with in the areas that they live and roam, such as their cages, coops, feed and water dishes, hay, plants, or soil. The ease with which the bacteria can spread makes the people caring for the poultry and those in their household susceptible to infection. Coming into contact with infected live poultry or any of the above items and then touching your mouth can result in infection. Keeping backyard poultry outside of the house and in their own living environment is a good way to ensure your own living area does not become contaminated. All cages and coops should be cleaned, as well as any materials or items used to clean them. Any clothing worn while caring for poultry should be kept out of the house, and hands should be washed with soap and water after coming into contact with poultry or their environment. If soap and water is not readily available, hand sanitizer should be used.

RESOURCES

Web Resources

Alcraft Egg Artistry, LLC	www.alcrafteggartistry.com
American Bantam Association	www.bantamclub.com
American Poultry Association	www.amerpoultryassn.com
The City Chicken	www.thecitychicken.com
The Eggery Place	www.theeggeryplace.com
International Art Guild	www.internationaleggartguild.com
Internet Center for Wildlife Damage Management	www.icwdm.com
The Livestock Conservancy	www.livestockconservancy.org
Metzer Farms	www.metzerfarms.com
Municipal Code Corporation	www.municode.com
My Pet Chicken	www.mypetchicken.com
Preventing Salmonella Infection	www.cdc.gov/healthypets/resources/salmonella-baby-poultry.pdf
Tractor Supply Company	www.tractorsupply.com
Ukrainian Gift Shop	www.ukrainiangiftshop.com
Uniquely Emu Products, Inc.	www.uniquelyemu.com

Publications

4-H Guide: Raising Chickens	Tara Kindschi
The American Standard of Perfection	American Poultry Association
The Chicken Health Handbook	Gail Damerow
City Chicks	Patricia Foreman
How to Raise Chickens	Christine Heinrichs
Poultry Press	www.poultrypress.com
Raising Poultry the Modern Way	Leonard S. Mercia
Your Chickens: A Kid's Guide to Raising and Showing	Gail Damerow

Mail Order Poultry Supplies

Brinsea	www.brinsea.com	
Crazy K Farm	www.crazykfarm.com	
Critter-Cages.com	www.critter-cages.com	
Cutler Supply (Pheasant, Poultry & Beekeeping)	www.cutlersupply.com	
Eggboxes.com	www.eggboxes.com	(800) 326-6667
First State Veterinary Supply	www.firststatevetsupply.com	
Fleming Outdoors	www.flemingoutdoors.com	(800) 624-4493
GQF Manufacturing Company Inc.	www.gqfmfg.com	
Ideal Poultry Breeding Farms, Inc.	www.idealpoultry.com	
Kemp's Koops	www.kempskoops.com	
Murray McMurray Hatchery	www.mcmurrayhatchery.com	
Seven Oaks Game Farm	www.poultrystuff.com	
Smith Poultry & Game Bird Supplies	www.poultrysupplies.com	

Organizations

American Livestock Breeds Conservancy	www.albc-usa.org
Association of Poultry Processors and Poultry Trade in the EU (AVEC)	www.avec-poultry.eu
Future Farmers of America (FFA)	www.ffa.org/home
4-H	www.4-h.org

About the Authors

Ronette Gehring, BVSc, MMedVet (Pharm), Dipl. ACVCP
is the Midwest Region Director for the Food Animal Residue Avoidance and Depletion program. She is an Associate Professor with the College of Veterinary Medicine at Kansas State University, where she studies the movement of drugs through the bodies of different animal species and how the resulting concentrations of drugs in tissues relate to their effect and residues in food-producing animals. Dr. Gehring has authored and coauthored many publications about drug residue avoidance in both the scientific and lay press.

Nancy Jefferson, Ph.D., Poultry Nutritionist
Nancy Jefferson is part of the nutrition and technical services team at a commercial feed company. She received her doctorate degree from West Virginia University, in 2008, and has worked in the feed industry for nine years. She currently lives on a farm in Ohio with her husband John and their young children. Together, they raise beef cattle and she keeps a small flock of poultry.

Brigid A. McCrea, PhD
is a native of California and grew up in both Los Angeles and Silicon Valley. She was introduced to farming through her local 4-H Club and began her lifelong love of chickens by participating in poultry shows. She entered the avian sciences program at the University of California, Davis, where she earned her bachelor of science degree. Her interest in poultry science and cooperative extension was furthered by her master of science work done with Dr. Joan Schrader, where her research examined *Salmonella* and *Campylobacter* recovery in chickens marketed as free-range or sold at Asian live fowl markets and correlated the prevalence to the management practices of the producers. After spending a year doing additional research on the microbiology of niche market poultry, she pursued her PhD in the poultry science program at Auburn University.

Her research program, under the advisement of Dr. Sacit Bilgili, was focused on *Salmonella*, *Campylobacter*, and *E. coli* recovery in broilers from the day of hatch through processing.

After completing her graduate program, she served as a postdoctoral researcher at the University of California, Davis, with Dr. Francine Bradley. She gained valuable extension experience by participating in the execution of the game fowl health assurance program as well as the poultry health inspection program. She took a position at Delaware State University in Dover, where she was assistant professor and extension poultry specialist. In that position, she had a primarily extension appointment, and her efforts were aimed at meeting the needs of small flock and niche market poultry producers. Her areas of interest include niche market poultry management, which includes organic, pastured, and heritage breed poultry production.

She has recently taken a position at Auburn University working as an extension specialist with 4-H Youth Development in the Alabama Cooperative Extension System. She is providing leadership over all the animal programs in the state as well as developing or revising curriculum to meet the needs of the programs. She enjoys working with the thousands of youth in the state that have an interest in, as well as participate in, animal programs, including programs such as 4-H Chick Chain.

Thinking of her roots in 4-H poultry, she also dedicates time participating in the annual National 4-H Poultry and Egg Conference. She also coordinates the Alabama 4-H Poultry Judging Contest, 4-H Golden Egg Contest, and 4-H Avian Bowl. She is also very dedicated to the transfer and application of biosecurity information to small and backyard flock owners through speaking engagements both nationally and internationally. She is coauthor of *The Chicken Whisperer's Guide to Keeping Chickens* and contributing author of *Chicken Fact or Chicken Poop*, a regular contributor to the "Plain Talk" column in *Chicken Whisperer Magazine*, and a regular guest speaker on the *Backyard Poultry with Andy the Chicken Whisperer* radio show.

Megin Nichols, DVM, MPH, DACVPM

serves as the enteric zoonoses activity lead at the Centers for Disease Control and Prevention (CDC). In this role, she works on multistate outbreaks of *Salmonella* and *E. coli* resulting from exposure to animals and pet products. Dr. Nichols has focused her work on investigating multistate outbreaks of human illness linked to petting zoos, small turtles, livestock with strains of multidrug-resistant *Salmonella*, and pet food products. In 2016, Dr. Nichols led the investigation of nine multistate outbreaks linked to live poultry in backyard flocks. A total of almost 900 people became ill in these outbreaks, the largest number of illnesses the CDC has recorded linked to live poultry. Prior to joining the CDC, Dr. Nichols worked as the principal investigator of the Active Bacteri

Prior to joining the CDC, Dr. Nichols worked as the principal investigator of the Active Bacterial Core Surveillance Program at the New Mexico Department of Health for five years. She received a bachelor of science degree in animal science from New Mexico State University, a doctor of veterinary medicine from Colorado State University, and a master of public health in food safety and biosecurity from the University of Minnesota. She was an Epidemiologic Intelligence Service (EIS) officer from 2008 to 2010 with the New Mexico Department of Health. Prior
to obtaining a DVM, she spent several years as a clinical veterinary assistant. Her areas of interest include zoonotic disease, food safety, and pediatric health.

Maurice Pitesky, DVM, MPVM, DACVPM
is a faculty member at the University of California Cooperative Extension (UCCE) with an appointment in poultry health and food safety epidemiology. Dr. Pitesky earned his bachelor of science in biology from UCLA, and his DVM and MPVM from the University of California, Davis. Dr. Pitesky is also boarded in preventative veterinary medicine (DACVPM).

Dr. Pitesky's research interests are focused in three major areas:

1. Using "traditional" epidemiological techniques and geographic information systems (GIS) and spatial statistics to understand how avian diseases move in time and space

2. Using traditional and "Next Generational" epidemiological and sequencing technologies to gain insights into *Salmonella* with respect to food safety

3. Gaining a better understanding of small-scale poultry production with respect to environmental sustainability, poultry heath, and food safety

Andy Schneider, "The Chicken Whisperer"
Andy Schneider, better known as "The Chicken Whisperer," has become the go-to guy for anything related to the backyard chicken movement. Over the years, he has helped a countless number of people start their very own backyard flocks. He is the national spokesperson for the USDA-APHIS Biosecurity for Birds program, editor in chief of *Chicken Whisperer Magazine*, host of the web radio show and podcast *Backyard Poultry with the Chicken Whisperer*, and founder of Fact or Chicken Poop?, a website that debunks information found on backyard-chicken-related blogs, forums, and books. He has started and organized some of the largest backyard chicken Meetup groups in the United States.

He has been featured on CNN, HLN, FOX, ABC, CBS, NBC, and NPR, as well as in the *Wall Street Journ*al, *Time* magazine, the *Economist*, *USA Today*, and countless other national and local publications. Andy travels with his family throughout the United States on Chicken Whisperer tours, educating people about the many benefits of keeping a small backyard flock of chickens. Andy says he's "spreading the chicken love from coast to coast!" An Atlanta native, Andy acquired his first chickens and turkeys almost twenty-five years ago, after moving to a rural area well north of the city. After moving back to the suburbs of Atlanta, he kept a small flock of urban chickens long before it was cool. You can visit his websites www.chickenwhisperer.com, www.factorchickenpoop.com, and www.chickenwhisperermagazine.com.

Zac Williams, Ph.D. Independent Poultry Scientist

Zac Williams was a former poultry science instructor at Tennessee Technological University (TTU). He served as poultry specialist for the state of Tennessee, addressing problems and educating farmers of both large- and small-scale poultry. While there, he laid the groundwork for the new poultry science program and research center at TTU, including funding and construction of a state-of-the-art multimillion dollar poultry research facility. He graduated with poultry science degrees from Mississippi State and Auburn universities. His background is primarily focused on waste management and diseases of poultry.

Index